The Patrick Moore Practical Astronomy Series

For further volumes:
http://www.springer.com/series/3192

Binocular Astronomy

Stephen Tonkin

Second Edition

Stephen Tonkin
Fordingbridge
Hampshire, UK

ISSN 1431-9756
ISBN 978-1-4614-7466-1   ISBN 978-1-4614-7467-8 (eBook)
DOI 10.1007/978-1-4614-7467-8
Springer New York Heidelberg Dordrecht London

Library of Congress Control Number: 2013941879

© Springer Science+Business Media New York 2007, 2014
This work is subject to copyright. All rights are reserved by the Publisher, whether the whole or part of the material is concerned, specifically the rights of translation, reprinting, reuse of illustrations, recitation, broadcasting, reproduction on microfilms or in any other physical way, and transmission or information storage and retrieval, electronic adaptation, computer software, or by similar or dissimilar methodology now known or hereafter developed. Exempted from this legal reservation are brief excerpts in connection with reviews or scholarly analysis or material supplied specifically for the purpose of being entered and executed on a computer system, for exclusive use by the purchaser of the work. Duplication of this publication or parts thereof is permitted only under the provisions of the Copyright Law of the Publisher's location, in its current version, and permission for use must always be obtained from Springer. Permissions for use may be obtained through RightsLink at the Copyright Clearance Center. Violations are liable to prosecution under the respective Copyright Law.

The use of general descriptive names, registered names, trademarks, service marks, etc. in this publication does not imply, even in the absence of a specific statement, that such names are exempt from the relevant protective laws and regulations and therefore free for general use.

While the advice and information in this book are believed to be true and accurate at the date of publication, neither the authors nor the editors nor the publisher can accept any legal responsibility for any errors or omissions that may be made. The publisher makes no warranty, express or implied, with respect to the material contained herein.

Springer is part of Springer Science+Business Media (www.springer.com)

Dedication

This book is for Ali and Bella.
May you never lose your fascination with the natural world.

Preface

Several years ago, a combination of age and abuse finally took its toll on my back and it became increasingly uncomfortable to use an equatorially mounted telescope for visual astronomy. I considered the option of setting up a system whereby I could operate a telescope remotely from the warmth and comfort of my study and see the resulting images on my computer screen. Almost immediately it became blindingly obvious to me that, while this is a pleasurable option for many amateur astronomers, it was not one that suited me. To do so would take me more into the realms of what some call "serious" amateur astronomy, which has rapidly embraced the advances that modern microelectronic technology has to offer, enabling the serious amateur to make significant contributions to astronomical knowledge. The thought of going further down this route brought it home to me: the reason that I "do" astronomy is for pleasure and relaxation and the option that I was considering was in danger of making it seem to me like another job.

I had always used binoculars for quick views of the sky when I did not have time to set up a telescope, and for more extended observing when, for example, I was waiting for a telescope to reach thermal equilibrium. I also always keep a binocular in my car so that I usually have an observing instrument reasonably close by. Now my back injury meant that my binoculars were the only astronomical instruments that I could comfortably use. I felt as if I was resigning myself to this. I considered my options again and decided that, if I was going to be "stuck" with binoculars, I might as well at least have some good-quality ones. Almost simultaneously, a large astronomical binocular was advertised for sale at an attractive price on *UK Astro Ads* and I took the opportunity and purchased it. This turned out to be the best decision of my astronomical life; it was like discovering visual astronomy all over again.

Why was this? Firstly, I became less "technological." I no longer had an equatorial mount to align, computers to set up, CCD camera to focus, power supplies to manage. Within minutes of making a decision to observe, I could be observing.

Secondly, and there are no other words for it, I was blown away by what I saw through the two eyepieces of a good 100 mm binocular. The first object that I turned my new acquisition to was the Great Nebula in Orion. It was like seeing it for the first time. I began to see detail that I had never before noticed visually, and some of this disappeared if I used only one eye. The pleasure of just sweeping the skies seeing what I can find is far greater than ever it was in a telescope—two eyes give one the impression that one is actually out there! Lastly, I found that I had stopped wondering if I would ever discover a comet or a supernova and had stopped thinking to myself that it was about time I did some more occultation timings. I was observing *purely* for pleasure. I realized that this was something that I had not done since I was a child. I had rediscovered my astronomical roots.

There can be pressure in amateur astronomical communities to participate in observing programs, to use one's hobby to advance the status of amateur astronomers. There can also be a tangible, and not always unspoken, attitude that someone who observes only, or even primarily, for pleasure does not really deserve to be called an amateur astronomer. My one regret is that it took me so many years to realize that this is a load of nonsense. The primary purpose of a hobby is enjoyment. If people find enjoyment in "serious" amateur astronomy, then all well and good, but, I contend, it is equally legitimate to enjoy it purely for recreation. Many have found that binoculars lead one to do exactly that.

Recreational observing is not the only application of binoculars; they are also well-suited to some aspects of serious astronomy. Big binoculars with their wide fields of view are excellent tools for visual comet hunters, as the names George Alcock and Yuji Hyakutaki attest. There are many variable star programs specifically for binoculars, such as that run by the *Society for Popular Astronomy*.

With even modest binoculars, there is sufficient in the sky to keep one enthralled for years; with good-quality big binoculars, there must be sufficient for decades. This book is for those who wish to explore that further, either with binoculars as an adjunct to a telescope or, as an increasing number of us are finding, as a main instrument. Its aim is to give a thorough understanding of the optical systems you will be using, and to indicate those criteria that should influence your choice of binocular. Once the choice is narrowed down, you will need to evaluate your options and there are simple tests you can do to give a good indication as to the potential of your choice. As with any aspect of astronomy, things do not stop with the optical system itself. You will, I hope, want to mount your binoculars—even small, normally hand-held, ones show so much more when mounted—and there are numerous accessories and techniques that can increase your observing comfort, pleasure, and efficacy. Lastly, of course, there are the objects themselves that you will observe. I have indicated which are suitable for small (50 mm aperture), medium (70 mm aperture), and large (100 mm aperture) binoculars. Obviously, all those in the 50 mm class are observable with a larger instrument (although a few are more pleasing with the wider field of the smaller instrument), and most of those for the larger instrument can at least be detected with the smaller one. These are intended as a "taster"—there are many more available to you. For example, some observers have seen all the Messier objects with 10×50 glasses! I hope that southern hemisphere readers will feel that I have made sufficient effort to include a good representation from their wonderful skies. The charts are simple black-on-white, as that is by far the easiest to read under red light.

Whatever category of binocular observer you fall into, there is something here for you. I hope you will get out there and find the same enjoyment from your binoculars as I continue to have from mine.

Fordingbridge, Hampshire, UK   Stephen Tonkin

Acknowledgements

The author's name is on the spine, but no book would exist without the inspiration and support, some of it unknowing, from a host of other people.

This book was gradually inspired over several years by a number of people. In addition to the numerous people with whom I have discussed binocular astronomy over the decades at astronomical meetings, star parties and on various Internet forums, those who deserve to be named are, in no particular order: Rob Hatch, who owned the first big binoculars through which I looked; Mike Wheatley, who showed me that 15×70s *can* be hand-holdable; Dave Strange, whose binocular chair (which looked—and felt—like some peculiar species of medieval torture instrument) got me thinking about the ergonomics of mounting systems; Bob Mizon, whose monthly Sky Notes at Wessex Astronomical Society meetings invariably exhorted people to use binoculars; Larry Patriarca of Universal Astronomics, whose superb binocular mounts make observing with binoculars a sheer pleasure; Bill Cook, who has talked opto-mechanical common sense to me for over a decade; Ed Zarenski, whose work inspired me to think more carefully about rigorously testing and evaluating binoculars; Konstantinos Makropoulos, who introduced me to some very efficient ways of checking binocular collimation; and Peter Drew, whose various ingenious binocular creations—two of which were stolen during an April 2012 break-in at the Astronomy Centre in Todmorden, West Yorkshire, UK—are sufficient to convert the most hardened one-eyed observer.

I am grateful to the following for permitting me to use their photographs: Florian Boyd, John Burns (of Strathspey Binoculars), Jim Burr (of Jim's Mobile Inc.), Norman Butler, Canon Inc., Jim Castoro (of The Binoscope Company), Chris Floyd (of Starchair Engineering Pty Ltd), Keith Harlow, Ted Ishikawa (of Hutech Inc), Axel Mellinger for his Milky Way Panorama, Gordon Nason, Bruce Sayre, Craig Simmons, and Rob Teeter (of Teeter's Telescopes). On the subject of photographs, thanks also go to my son, Tim Tonkin, for whom the consequence of a childhood of being taught how to hold binoculars properly was to model the holds for the relevant photographs.

The charts in Part 2 were prepared using Bill Gray's superb *Guide v9*, from http://projectpluto.com and the binocular view simulations in Chaps. 3 and 9 were based on the output of the freeware planetarium program *Stellarium* (http://stellarium.org). The output of both of these programs is reproduced (as amended) here under GNU Public License.

Among those at Springer whom I must thank are John Watson for his continuing support, Maury Solomon for her encouragement and support with initiating this 2nd edition, and Nora Rawn who guided me through the production process. The book was produced by SPi Technologies. Thanks are due to their editorial team and especially to Mrs Indumathy Saikumar who converted my sometimes convoluted British English into a more readable American version of the same language, and who picked up numerous silly grammatical and typographical errors in the text; any that remain are my responsibility.

Finally, there is my wife, Louise Tonkin, whose support ranges from gentle encouragement to a tolerance of the socially inconvenient times that I choose to spend writing, and which is punctuated by regular cups of strong espresso!

About the Author

Stephen Tonkin, B.Sc. (Hons), F.R.A.S., has been a keen amateur astronomer since childhood and now spends most of his time doing astronomical education and outreach, both as a Lecturer in Astronomy for an adult education college, and independently with his own organization, *The Astronomical Unit*. He organizes and leads astronomy courses and talks, public observing, and astronomy-related storytelling for children and adults. In 2000, he was elected as a Fellow of the Royal Astronomical Society.

He has been using binoculars for astronomy for over 40 years, initially under the pristine African skies under which he grew up, and as his main observing instrument for the last decade. He actively promotes and encourages the use of binoculars within the amateur astronomy community and publishes a monthly e-zine, *The Binocular Sky Newsletter*, for binocular astronomers. He also writes the monthly *Binocular Tour* in *Sky at Night* magazine.

He now lives on the edge of the New Forest, which has some of the darkest skies in southern England. On clear moonless nights when he's not working, he can usually be found at one of these dark sites, exploring the night sky with his binoculars.

Contents

Part I Binoculars

1 Why Binoculars? 3
    Portability 4
    Ease of Setup 4
    The Binocular Advantage 5
    The 5-mm Exit Pupil 6
    Small Focal Ratio and Aberrations 7
    Conclusion 8
    Bibliography 8

2 Binocular Optics and Mechanics 9
    Objective Lens Assemblies 11
    Eyepieces 11
    Prisms 12
    Coatings 24
    Aberrations 29
    Aperture Stops and Vignetting 35
    Focusing Mechanisms 36
        Center Focus (Porro Prism) 36
        Center Focus (Roof Prism) 36
        Independent Focus 36
    Collimation 37
    Bibliography 40

3 Choosing Binoculars ... 43
Deciding What You Need ... 43
Binocular Specifications ... 44
What Size? ... 46
Field of View ... 48
Eye Relief ... 50
Handheld Binoculars ... 53
Mounted Binoculars ... 55
   Budget Versus Quality ... 57
Binoviewers ... 58
Zoom Binoculars ... 61
Bibliography ... 61

4 Evaluating Binoculars ... 63
Preliminary Tests ... 64
Field Tests ... 73
Additional Tests for Used Binoculars ... 76

5 Care and Maintenance of Binoculars ... 77
Rain Guards ... 78
Storage ... 78
Desiccants ... 79
Grit ... 80
Cleaning ... 80
Dismantling Binoculars ... 82
Right Eyepiece Diopter Adjustment ... 89
The Solution ... 89
Collimation ... 91
Bibliography ... 94

6 Holding and Mounting Binoculars ... 95
Hand-Holding ... 95
"Informal" Supports ... 99
Mounting Brackets ... 100
Monopods ... 103
Neckpod ... 104
Bodge-o-pod ... 106
Photo Tripods ... 107
Fork Mounts ... 109
Mirror Mounts ... 110
Parallelogram Mounts ... 111
Observing Chairs ... 113
Summary ... 116
Bibliography ... 117

Contents   xvii

7   Binocular Telescopes 119
   Binocular Telescopes 119

8   Observing Accessories 129
   Finders 129
   Filters 132
   Dew Prevention and Removal 133
   Compass 135
   Charts and Charting Software 135
   Torches (Flashlights) 137
   Storage and Transport Container 138
   Software Sources 139

9   Observing Techniques 141
   Personal Comfort 141
   Observing Sites 143
   Observing Techniques 144

Part II Deep Sky Objects for Binoculars

10 Overview 149
   The Object Catalogues 150
      Summary Charts 151
   North Polar Region 152
   North RA 22 h 30 m to 01 h 30 m 153
   South RA 22 h 30 m to 01 h 30 m 154
   North RA 01 h 30 m to 04 h 30 m 155
   South RA 01 h 30 m to 04 h 30 m 156
   North RA 04 h 30 m to 07 h 30 m 157
   South RA 04 h 30 m to 07 h 30 m 158
   North RA 07 h 30 m to 10 h 30 m 159
   South RA 07 h 30 m to 10 h 30 m 160
   North RA 10 h 30 m to 13 h 30 m 161
   South RA 10 h 30 m to 13 h 30 m 162
   North RA 13 h 30 m to 16 h 30 m 163
   South RA 13 h 30 m to 16 h 30 m 164
   North RA 16 h 30 m to 19 h 30 m 165
   South RA 16 h 30 m to 19 h 30 m 166
   North RA 19 h 30 m to 22 h 30 m 167
   South RA 19 h 30 m to 22 h 30 m 168
   South Polar Region 169
   Objects by Type (Listed in Order of Right Ascension) 169
      Asterisms 169
      Dark Nebulae 170
      Emission Nebulae 170

- Galaxies 170
- Globular Clusters 171
- Multiple Stars 172
- Open Clusters 173
- Planetary Nebulae 175
- Reflection Nebulae 175
- Supernova Remnants 175
- Nearby Star 175
- Variable Stars 175

Objects by Binocular Aperture (Listed in Order of Right Ascension) 176

Objects by Constellation 181
- Andromeda 181
- Aquarius 181
- Aquila 181
- Ara 182
- Aries 182
- Auriga 182
- Boötes 182
- Camelopardalis 182
- Cancer 182
- Canis Major 183
- Carina 183
- Cassiopeia 183
- Centaurus 183
- Cepheus 184
- Cetus 184
- Coma 184
- Corona Australis 184
- Corvus 184
- Crux 185
- Canes Venatici 185
- Cygnus 185
- Delphinus 185
- Dorado 185
- Draco 186
- Eridanus 186
- Gemini 186
- Hercules 186
- Hydra 186
- Lacerta 186
- Leo 186
- Lepus 187
- Monoceros 187
- Norma 187
- Ophiuchus 187

	Orion	188
	Pavo	188
	Pegasus	188
	Perseus	188
	Pictor	188
	Puppis	189
	Sagitta	189
	Sagittarius	189
	Scorpius	189
	Sculptor	190
	Scutum	190
	Serpens	190
	Sextans	190
	Taurus	190
	Telescopium	191
	Triangulum	191
	Triangulum Australis	191
	Tucana	191
	Ursa Major	191
	Ursa Minor	191
	Vela	192
	Virgo	192
	Vulpecula	192
	Bibliography	192
11	**December Solstice to March Equinox (RA 04:00 h to 10:00 h)**	**193**
	Perseus: Emission Nebula: NGC 1499 (the *California Nebula*) (70 mm)	194
	Perseus: Open Cluster: NGC 1528 (70 mm)	195
	Eridanus: Planetary Nebula: NGC 1535 (100 mm)	196
	Taurus: Open Cluster: Melotte 25 (C41, the *Hyades*) (50 mm)	197
	Taurus: Open Cluster: NGC1647 (70 mm)	198
	Taurus: Open Cluster: NGC 1746 (70 mm)	199
	Taurus: Supernova Remnant: M1 (NGC 1952, the Crab Nebula) (100 mm)	200
	Lepus: Variable Star: R Leporis (*Hind's Crimson Star*) (70 mm)	201
	Lepus: Double Star: γ Leporis (50 mm)	202
	Auriga: Asterism: The *Leaping Minnow* (50 mm)	203
	Auriga: Three Open Clusters: M36 (NGC 1960), M37 (NGC 2099), and M38 (NGC 1912) (70 mm)	204
	Dorado: Galaxy and Emission Nebula: *Large Magellanic Cloud* and NGC 2070 (C103, *Tarantula Nebula, Loop Nebula,* 30 Doradus) (100 mm)	205
	Pictor: Double Star: θ Pictoris (100 mm)	206
	Orion: Open Cluster: Collinder 65 (50 mm)	207

Orion: Nebulosity and Clusters: M42 (NGC 1976), M43 (NGC 1982),
NGC 1973, 1975, 1977, and 1980 (50 mm) 208
Orion: Open Cluster: Cr 70 (50 mm) 210
Orion: Multiple Star: σ Orionis (50 mm) 211
Orion: Nebula: NGC 2024 (*the Flame Nebula, the Burning Bush,
the Ghost of Alnitak*) (70 mm) 212
Orion: Emission Nebula: M78 (NGC 2068) (70 mm) 213
Gemini: Open Cluster: M35 (NGC 2168) (50 mm) 214
Monoceros: Open Cluster: NGC 2239 (NGC 2244, C50) (70 mm) 215
Monoceros: Open Cluster: NGC 2264 (the *Christmas
Tree Cluster*) (70 mm) 216
Monoceros: Open Cluster: M50 (NGC 2323) (50 mm) 217
Monoceros: Open Cluster: NGC 2353 (100 mm) 218
Canis Major: Open Cluster: M41 (NGC 2287) (50 mm) 219
Canis Major: Open Cluster: NGC 2362 (C64) (100 mm) 220
Puppis: Open Clusters: M46 (NGC 2437) and M47
(NGC 2422) (50 mm) 221
Camelopardalis: Galaxy: NGC 2403 (C7) (100 mm) 222
Carina: Open Cluster: NGC 2516 (C96) (100 mm) 223
Vela: Open Cluster: NGC 2547 (100 mm) 224
Puppis: Open Cluster: NGC 2539 (100 mm) 225
Puppis: Open Cluster: M93 (NGC 2447) (70 mm) 226
Puppis: Open Cluster: NGC 2451 (50 mm) 227
Puppis: Open Cluster: NGC 2477 (C71) (70 mm) 228
Puppis: Open Cluster: NGC 2546 (100 mm) 229
Hydra: Open Cluster: M48 (NGC 2548) (70 mm) 230
Vela: Open Cluster: IC 2391 (C85, the *Omicron Velorum Cluster*)
(50 mm) 231
Cancer: Open Cluster: M44 (NGC 2632, *Praesepe*,
the *Beehive Cluster*) (50 mm) 232
Cancer: Open Cluster: M67 (NGC 2682) (70 mm) 233
Sextans: Double Star: 9 Sextantis (100 mm) 234
Ursa Major: Galaxy Pair: M81 (NGC 3031) and M82
(NGC 3034) (100 mm) 235

12 March Equinox to June Solstice (RA 10:00 h to 16:00 h) 237
Carina: Open Cluster: NGC 3114 (50 mm) 238
Sextans: Galaxy: NGC 3115 (C53, the *Spindle Galaxy*) (100 mm) 239
Hydra: Planetary Nebula: NGC 3242 (C59, the *Ghost
of Jupiter*) (100mm) 240
Carina: Open Cluster: IC 2602 (C102, the θ *Carinae Cluster*,
the *Southern Pleiades*) (50 mm) 241
Carina: Emission Nebula: NGC 3372 (C92, η *Carinae Nebula*)
(50 mm) 242

Contents xxi

Leo: Galaxy Trio: M95 (NGC 3351), M96 (NGC 3368), and M105 (NGC 3379) (100 mm)	243
Leo: Galaxy: NGC 3521 (100 mm)	244
Leo: Galaxy: NGC 3607 (100 mm)	245
Leo: Galaxy Trio: M65 (NGC 3623), M66 (NGC 3627) and NGC 3628 (100 mm)	246
Ursa Major: Planetary Nebula: M97 (NGC 3587, the *Owl Nebula*) (100 mm)	247
Ursa Major: Asterism: M40 (100 mm)	248
Corvus: Planetary Nebula: NGC 4361 (100 mm)	249
Centaurus: Open Cluster: NGC 3766 (C97, the *Pearl Cluster*) (100 mm)	250
Centaurus: Open Cluster and Supernova Remnant: IC 2944 (C100, *the Running Chicken, the λ Centauri Nebula*) (100 mm)	251
Canes Venatici: Galaxy: M106 (NGC 4258) (100 mm)	252
Canes Venatici: Galaxy Pair: NGC 4631 (C32, *the Whale Galaxy*) and NGC 4656 (100 mm)	253
Canes Venatici: Carbon Star: Y CVn (*La Superba*) (50 mm)	254
Canes Venatici: Galaxy: M94 (NGC 4736) (70 mm)	255
Canes Venatici: Galaxy: M63 (NGC 5055, the *Sunflower Galaxy*) (70 mm)	256
Canes Venatici: Galaxy: M51 (NGC 5194, the *Whirlpool Galaxy*) (100 mm)	257
Canes Venatici: Globular Cluster: M3 (NGC 5272) (70 mm)	258
Coma Berenices: Open Cluster: Melotte 111 (50 mm)	259
Coma Berenices: Galaxy: NGC 4559 (C36) (100 mm)	260
Coma Berenices: Galaxy: NGC 4565 (C38, *Berenice's Hair Clip, the Needle Galaxy*) (100 mm)	261
Coma Berenices: Galaxy: M64 (NGC 4826, the *Black Eye Galaxy*) (70 mm)	262
Coma Berenices: Globular Cluster: M53 (NGC 5024) (100 mm)	263
Musca: Globular Cluster: NGC 4372 (C108) (100 mm)	264
Musca: Globular Cluster: NGC 4833 (C105) (100 mm)	265
Crux: Open Cluster: NGC 4755 (C94, the *Jewel Box*) (50 mm)	266
Virgo: Galaxy Chain: NGC 4374 (M84), 4406 (M86), 4438, 4473, 4477, and 4459 (*Markarian's Chain*) (100 mm)	267
Virgo: Galaxy: M49 (NGC 4472) (70 mm)	268
Virgo: Galaxy Group: M87 (NGC 4486) and Friends (70 mm)	269
Virgo: Galaxy Pair: M59 (NGC 4621) and M60 (NGC 4649) (70 mm)	270
Virgo: Galaxy: M104 (NGC 4594, the *Sombrero Galaxy*) (100 mm)	271
Hydra: M68 (NGC 4590) (100 mm)	272
Hydra: Galaxy: M83 (NGC 5263) (100 mm)	273
Centaurus: Galaxy: NGC 5128 (C77, *Centaurus A*) (100 mm)	274
Centaurus: Globular Cluster: NGC 5139 (C80, Omega Centauri) (50 mm)	275

Ursa Major: Galaxy: M101 (NGC 5457) (100 mm) 276
Draco: Galaxy: NGC 5866 (100 mm) 277
Draco: Galaxy: NGC 5907 (the *Splinter Galaxy*) (100 mm) 278
Boötes: Variable Star: RV Boötis (100 mm) 279
Boötes: Multiple Stars: δ Boötis and 50 Boötis (100 mm) 280
Serpens: Globular Cluster: M5 (NGC 5904) (70 mm) 281

13 June Solstice to September Equinox (RA 16:00 h to 22:00 h) 283

Triangulum Australe: Open Cluster: NGC 2065 (100 mm) 284
Norma: Open Cluster: NGC 6067 (100 mm) 285
Scorpius: Globular Clusters: M4 (NGC 6121)
  and NGC 6144 (70 mm) 286
Scorpius: Open Cluster: NGC 6231 (C76) (50 mm) 287
Scorpius: Open Cluster: NGC 6322 (100 mm) 288
Scorpius: Open Cluster: M6 (NGC 6405, the *Butterfly
  Cluster*) (50 mm) 289
Scorpius: Open Cluster: M7 (NGC 6475, *Ptolemy's
  Cluster*) (50 mm) 290
Ophiuchus: Triple Star: ρ Ophiuchi (100 mm) 291
Ophiuchus: M12 (NGC 6218) (70 mm) 292
Ophiuchus: M10 (NGC 6254) (70 mm) 293
Ophiuchus: M62 (NGC 6266) (100 mm) 294
Ophiuchus: M19 (NGC 6273) (70 mm) 295
Ophiuchus: M14 (NGC 6402) (70 mm) 296
Ophiuchus: Open Cluster: IC 4665 (*the Summer Beehive*)
  (70 mm) 297
Ophiuchus: Star: *Barnard's Star* (70 mm) 298
Ophiuchus: Open Cluster: Melotte 186 (50 mm) 299
Ophiuchus: Planetary Nebula: NGC 6572 (100 mm) 300
Ophiuchus: Open Cluster: NGC 6633 (100 mm) 301
Hercules: Globular Cluster: M13 (NGC 6205) (50 mm) 302
Hercules: Globular Cluster: M92 (NGC 6341) (100 mm) 303
Ara: Globular Cluster: NGC 6397 (C86) (100 mm) 304
Corona Australis: Globular Clusters: NGC 6541 (C78)
  and NGC 6496 (100 mm) 305
Sagittarius: Open Cluster: M23 (NGC 6494) (70 mm) 306
Sagittarius: Emission Nebula: M20 (NGC 6514,
  the *Trifid Nebula*) (100 mm) 307
Sagittarius: Open Cluster and Nebulosity: NGC 6530 and M8
  (NGC 6523, the *Lagoon Nebula*) (50 mm) 308
Sagittarius: Star Cloud: M24 (50 mm) 309
Sagittarius: Open Cluster: M18 (NGC 6613) (100 mm) 310
Sagittarius: Emission Nebula: M17 (NGC 6618, the Omega
  Nebula or Swan Nebula) (100 mm) 311

Sagittarius: Globular Cluster: M28 (NGC 6626) (70 mm)	312
Sagittarius: Open Cluster: M25 (IC 4725) (100 mm)	313
Sagittarius: Globular Cluster: M22 (NGC 6656) (70 mm)	314
Sagittarius: Globular Cluster: M54 (NGC 6715) (100 mm)	315
Sagittarius: Globular Cluster: NGC 6723 (100 mm)	316
Sagittarius: Globular Cluster: M55 (NGC 6809) (70 mm)	317
Telescopium: Globular Cluster: NGC 6584 (100 mm)	318
Serpens: Emission Nebula and Cluster: M16 (NGC 6611, the *Eagle Nebula*) (100 mm)	319
Serpens: Open Cluster: IC 4756 (50 mm)	320
Serpens: Double Star: θ Serpentis (100 mm)	321
Scutum: Open Cluster: M26 (NGC 6694) (70 mm)	322
Scutum: Open Cluster: M11 (NGC 6705, *Wild Duck Cluster*) (50 mm)	323
Scutum: Globular Cluster: NGC 6712 (100 mm)	324
Pavo: Globular Cluster: NGC 6752 (C 93) (100 mm)	325
Aquila: Open Cluster: NGC6709 (100 mm)	326
Aquila: Open Cluster: NGC 6738 (100 mm)	327
Aquila: Planetary Nebula: NGC 6781 (100 mm)	328
Aquila: Dark Nebulae: Barnard 142, 143 (*Barnard's E*) (70 mm)	329
Vulpecula: Asterism: (Cr 399, *Brocchi's Cluster*, the *Coathanger*) (50 mm)	330
Vulpecula: Planetary Nebula: M27 (NGC 6853, the *Dumbbell Nebula*) (50 mm)	331
Sagitta: Double Star: ε Sagittae (100 mm)	332
Sagitta: Cluster: M71 (NGC 6838) (100 mm)	333
Cygnus: Double Star: β Cyg (*Albireo*) (50 mm)	334
Cygnus: Open Cluster: M29 (NGC 6913) (70 mm)	335
Cygnus: Dark Nebula: LDN 906 (B 348, the *Northern Coalsack*) (50 mm)	336
Cygnus: Supernova Remnant: *Veil Nebula* NGC 6960 (C34), NGC 6992 (C33) and 6995 (100 mm)	337
Cygnus: Emission Nebula: NGC 7000 (C20, the *North American Nebula*) (50 mm)	338
Cygnus: Double Star: 61 Cygni (70 mm)	339
Cygnus: Open Cluster: M39 (NGC 7092) (70 mm)	340
Delphinus: Globular Cluster: NGC 6934 (C47) (100 mm)	341
Pegasus: Globular Cluster: M15 (NGC 7078) (50 mm)	342
Aquarius: Globular Cluster: M2 (NGC 7089) (50 mm)	343
Aquarius: Double Star: Struve 2809 (100 mm)	344
Cepheus: Open Cluster: IC1396 (50 mm)	345
Cepheus: Red Giant: μ Cep (the *Garnet Star*) (50 mm)	346

14 September Equinox to December Solstice (RA 22:00 h to 04:00 h) 347

Lacerta: Open Cluster: NGC 7209 (70 mm) 348
Lacerta: Open Cluster: NGC 7243 (70 mm) 349
Cepheus: Open Cluster: NGC 7235 (70 mm) 350
Cepheus: Open Cluster: NGC 7510 (70 mm) 351
Aquarius: Planetary Nebula: NGC 7293 (C63, the *Helix Nebula*) (100 mm) 352
Sculptor: Galaxy: NGC 55 (C72) (100 mm) 353
Sculptor: Galaxy and Globular Cluster : NGC 253 (C65) and NGC 288 (70 mm) 354
Sculptor: Galaxy: NGC 300 (C70) (100 mm) 355
Vela: Open Cluster: NGC 3228 (100 mm) 356
Tucana: Globular Cluster: NGC 104 (C106, 47 Tucanae) (100 mm) 357
Tucana: Galaxy: NGC 292 (*Small Magellanic Cloud*) (50 mm) 358
Andromeda: Galaxy: M31 (NGC 224, the *Great Andromeda Galaxy*) (50 mm) 359
Andromeda: Open Cluster and Double Star: NGC 752 (C28) and 56 And (70 mm) 360
Cetus: Galaxy: NGC 247 (C62) (100 mm) 361
Pisces: Double Star: ψ^1 Piscium (100 mm) 362
Pisces: Double Star: ζ Piscium (100 mm) 363
Andromeda: Open Cluster: NGC 7686 (70 mm) 364
Cassiopeia: Open Cluster: Stock 12 (70 mm) 365
Cassiopeia: Open Cluster: M52 (NGC 7654) (100 mm) 366
Cassiopeia: Open Cluster: NGC 7789 (70 mm) 367
Cassiopeia: Open Cluster: NGC 225 (70 mm) 368
Cassiopeia: Open Cluster: NGC 436 (100 mm) 369
Cassiopeia: Open Cluster: NGC 457 (C13) (the *ET Cluster*, the *Owl Cluster*) (100 mm) 370
Cassiopeia: Open Cluster: NGC 663 (C10) (50 mm) 371
Cassiopeia: Open Cluster: NGC 654 (70 mm) 372
Cassiopeia: Open Cluster: Cr 463 (70 mm) 373
Cassiopeia: Open Clusters: Mel 15 and NGC 1027 (70 mm) 374
Camelopardalis: Open Cluster: Stock 23 (70 mm) 375
Andromeda: Open Cluster: NGC 956 (100 mm) 376
Triangulum: Galaxy: M33 (NGC 598, the *Pinwheel Galaxy*) (50 mm) 377
Aries: Triple Star: 14 Arietis (50 mm) 378
Eridanus: Galaxy: NGC 1232 (100 mm) 379
Cetus: Variable Star: o Ceti (*Mira*) (50 mm) 380
Cetus: Galaxy: M77 (NGC 1068) (100 mm) 381
Cassiopeia: Open Cluster: Stock 2 (the *Muscleman Cluster*) (70 mm) 382

Perseus: Open Clusters: NGC 884 and NGC 869
(C14, the *Double Cluster*) (50 mm) 383
Perseus: Open Cluster: M34 (NGC 1039) (50 mm) 384
Perseus: Open Cluster: Melotte 20 (Cr 39, the *Alpha Persei
Moving Cluster*) (50 mm) 385
Perseus: Open Cluster: NGC 1342 (70 mm) 386
Ursa Minor: Asterism: The *Engagement Ring* (70 mm) 387
Taurus: Open Cluster: M45 (the *Pleiades*) (50 mm) 388
Camelopardalis: Asterism: *Kemble's Cascade* (70 mm) 389

Appendix 1 391

Appendix 2 397

Appendix 3 403

Appendix 4 411

Appendix 5 417

Appendix 6 419

Appendix 7 421

Index 429

Part I
Binoculars

Chapter 1

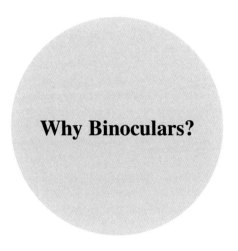

Why Binoculars?

Amateur astronomers usually view small- and medium-aperture (50–70 mm) binoculars either as an inexpensive "entry level" instrument to the hobby or as a useful accessory to a more experienced observer's "main" instrument, a telescope. There is a great deal of justification for this. Binoculars do indeed make excellent starter instruments for new observers, especially those of limited financial means. A medium-aperture binocular of reasonable quality is not only less expensive than the cheapest useful astronomical telescopes, but it is also much more intuitive to use, easier to set up, more portable, and has more obvious uses outside astronomy, for example, bird-watching or horse racing. It also enables the new observer to engage in useful observing programs, such as the Society for Popular Astronomy's variable star program.[1]

Where the more experienced observer is concerned, the wider field of a binocular is ideal for having a preliminary scan around the sky in order to evaluate it at the beginning of an observing session and is also useful in conjunction with the telescope's finder as an aid to hunting the objects to be observed. Additionally, there are large objects with low surface brightness, such as the Pinwheel Galaxy (M33, NGC 598), that are distinctly easier to see in such binoculars than they are in most telescopes of even twice the aperture.

What an increasing number of experienced observers are coming to realize is that the binocular is not limited to being an adjunct to a telescope, but is an exceptionally valuable astronomical instrument in its own right. Many of the advantages of the binocular when used for its "beginner" or "adjunct" purpose translate to its advanced use.

---

[1] See http://www.popastro.com

Portability

There are two facets to the portability of binoculars. The first is the compactness and weight of the instrument itself. A 10×50 binocular is possibly the most common starting binocular and adjunct binocular. It is typically about 18 cm (7 in.) long and about the same width and usually weighs a kilogram (2.2 lbs) or less, considerably less in the case of lightweight models. Secondly, binoculars of this size and weight can easily be handheld for moderate periods of time, so they do not need a mount to be carried with them. Even 15×70 or 16×70 binoculars, which are typically about 28 cm (11 in.) long and between 1.2 and 2.5 kg (2.6 and 5.5 lbs) in weight, may be handheld for short periods.

Of course, all binoculars will benefit from being mounted. If a mount is to be carried with this size of binocular, a reasonably sturdy photographic monopod or tripod with a pan/tilt head or, better, a trigger-grip ball-head will suffice for binoculars up to 80 mm in aperture, or 100 mm if they are the lighter-weight ones. However, it does need to be stated at the outset that the photographic tripod with pan/tilt head, although commonly used, is far from ideal as a binocular mount for astronomy (see Chap. 6).

Ease of Setup

Binoculars of 100-mm aperture or smaller are usually trivially easy to set up. If they are to be handheld (usually 50 mm or smaller), all that is required is that the interpupillary distance and focus are set. Unlike large telescopes, they do not normally require time to reach thermal equilibrium, so they can literally be regarded as "grab-and-go" instruments, with observations being made within a minute or so of the decision to observe!

Even larger binoculars are generally considerably simpler to set up than many telescopes. Binoculars are not generally equatorially mounted on account of the awkward positions that such mounting would require of the observer's head! For this reason, binoculars are usually mounted on some form of altazimuth mount, often a photographic tripod and head. Even with 6-kg (13.5-lb) binoculars on a sophisticated parallelogram mount, I routinely find that I am observing in less than 10 min of having made the decision to observe.

The Binocular Advantage

It is generally acknowledged, and empirical experiments confirm, that, using two eyes, our threshold of detection of faint objects is approximately 1.4 times as good as with one eye.[2] This is a consequence of what is called *binocular summation*,[3] which is itself probably a result of at least two different phenomena:

- *Statistical summation.* For objects of a low threshold of visibility, there is a greater probability that photons from the object will be detected by at least one of two detectors (in this case, eyes) than by a single detector. If the probability of detection in one detector is just over 0.5, then the probability of detection in both is indeed approximately 1.4 times greater; e.g., for a detection probability of 0.6 in one detector, the probability of detection in one of two identical detectors is given by:

$$P(\text{Both}) = P(\text{Right}) \textbf{ OR } P(\text{Left}) - (P(\text{Right}) \textbf{ AND } P(\text{Left}))$$
$$= 0.6 + 0.6 - (0.6 \times 0.6)$$
$$= 0.84$$
$$0.84 / 0.6 = 1.4$$

- *Physiological summation.* This is essentially an improvement of signal-to-noise ratio (SNR). The signals from each eye are added, but the random neural noise is partially cancelled. If the noise is random, the resulting improvement in SNR will be $\sqrt{2}$, i.e., approximately 1.4.

The consequence of binocular summation is that, with two eyes, we experience an improvement both in acuity of vision and in contrast. This is apparent when we have our eyes tested by an optometrist, where we notice that the eye chart is easier to read with both eyes than with one eye alone. It is easy to demonstrate this with binoculars: find an object that you can only just detect, or a double star that you can only just split, with both eyes, and then cap each objective in turn. You may even notice it while reading this page! However, this is only true for well-corrected vision; if the image in one eye is sufficiently degraded, then the consequence is that binocular vision is degraded to below the performance for the good eye. This obviously has implications for when we use binoculars.

Another bonus of using two eyes is stereopsis. Although astronomical objects are obviously far too distant for them to be seen with true stereoscopic vision, when

---

[2] For example, Dickinson & Dyer, 1991, p.26; Harrington, 1990, p2; Salmon
[3] For example, Salmon (ibid)

we use both eyes, there is an illusion of stereoscopic vision that enhances the aesthetic attributes of many objects. I find this effect particularly apparent with rich open clusters, especially when there are stars of obviously different colors.

Lastly, when you observe with two eyes, one of them sees the small part of the field of the other eye that is obliterated by the blind spot, the location on the retina where the optic nerve enters the eye. In this sense, the binocular can be said to give a more complete view than single-eye observing.

The 5-mm Exit Pupil

There is a lot of "internet wisdom" that suggests that the ideal exit pupil for binocular astronomy is 5 mm. This is based on a lot of assumptions, some of which (e.g., the change of pupil size with age—see below) are incorrect. Most binoculars for astronomy will give an exit pupil in the region of 3–5 mm. There are obvious exceptions to this. There are occasional "fashions" for using exit pupils of up to 7 mm in both medium (e.g., 7×50) and giant (e.g., 15×110, 25×150) binoculars, but there are very good reasons not to do so, as only a few objects can benefit from this even if our eyes' pupils do dilate that much. Similarly, there are some larger astronomical binoculars, usually with interchangeable eyepieces, where the exit pupil is smaller than 3 mm.

The change of pupil size with age is one bone of contention. Conventional wisdom dictates that, by the age of 40 years, the dark-adapted pupil diameter (DAPD) is limited to 5 mm. This is clearly an incorrect generalization. It may be true for some individuals, but it is certainly not true for all. At over two decades older than the conventional "5-mm age," my pupils both open to more than 6 mm, and a recent study[4] has demonstrated that the average DAPD does not fall to 5 mm until after the age of 79!

However, there are still distinct advantages in using an exit pupil in the 2.5–5-mm range. In no particular order they are:

- There is sufficient brightness to see most of the extended objects that are visible with a larger exit pupil. (Notable exceptions are the *Pinwheel Galaxy* (M33) and the *North American Nebula* (NGC 7000), both of which are better with a larger exit pupil, if our eyes can accommodate it.)
- Most observers' pupils do not dilate much beyond 6.5 mm. The eye's pupil therefore vignettes the light from the binocular if the exit pupil is larger than 6.5 mm.
- It is easier to position the eyes so that the entire exit pupil is contained by the eye's pupil if the exit pupil is smaller than the eye's pupil.
- Aberrations in the eye's lens and cornea tend, as they do in the lenses of optical instruments, to be more severe towards the periphery of the pupil than they do

---

[4]Bradley et al., 2011

at the center. Many normally bespectacled observers find that they can, with smaller exit pupils, observe satisfactorily without spectacles.
- Larger exit pupils imply lower magnification. Most binocular objects are easier to resolve with greater magnification and many are easier to identify. An object is fully resolved on the retina when the exit pupil is about 1 mm, although this is impracticably small for binoculars.
- The higher magnification results in greater contrast, on account of the sky itself being an extended object and consequently dimmed by greater magnification.
- Smaller exit pupils imply smaller real fields of view, so lateral chromatic aberration is reduced.

The obvious disadvantages are:

- Extended objects are fainter than they are with a larger exit pupil, assuming the eye can accommodate the larger pupil.
- Larger exit pupils imply lower magnifications, with consequently more relaxed tolerances for collimation between the tubes.

As with so many things in observational astronomy, there is a matter of preference. For a small handheld astronomical binocular, an exit pupil of 4–5 mm (e.g., 10×42, 10×50, 15×70) offers a good compromise between having sufficient magnification to darken the background sky and enhance contrast on the one hand and a large enough exit pupil to give a bright image on the other hand. For a mounted binocular, even dropping below 3 mm can be advantageous. My 100-mm binocular offers the option of ×20 (5-mm exit pupil) and ×37 (2.7-mm exit pupil); at the time of writing, it is over 4 years since I have used the ×20 eyepieces; such is the benefit of the higher magnification and the convenience of my not needing spectacles to observe with a 2.7-mm exit pupil.

Small Focal Ratio and Aberrations

Most binoculars have objectives that operate at around f/3.5 to f/5, although there are some specialist astronomical binoculars, intended for use at relatively high magnification, that have greater focal ratios.

Most optical aberrations are exacerbated with "fast" (i.e., low focal ratio, thus photographically "fast") objectives. For a normal achromatic doublet that does not use exotic glasses, the rule of thumb is that the focal ratio must be no less than three times the diameter of the aperture, measured in inches (1 in. = 25.4 mm), for axial (longitudinal) color correction to be acceptable. This is equivalent to stating that a 50-mm objective must work at f/6 and a 100 mm at f/12 or that the limit for f/5 is 42 mm. This latter equivalent is a reason for the good reputation for optical quality of many 42-mm binoculars. If optical quality is to be maintained at greater apertures without a concomitant increase in focal ratio, either expensive exotic glasses or extra lens elements or both must be employed. Several modern specialist astronomical binoculars have slower f-ratios, some as low as f/7.5 or f/8. An example of this is the Takahashi Astronomer, a 22×60 specialist astronomical binocular, which uses a combination of exotic (fluorite) glass and a focal ratio of 5.9 to give some of the crispest and most contrasty images that I have seen in an astronomical binocular.

Lower focal ratios have light cones that are more obtuse, and obtuse light cones are more demanding of eyepiece quality than are those that are more acute. This means that, for image quality to be preserved, higher-quality eyepieces are needed and thus greater expense is required.

Conclusion

Binoculars offer a relatively inexpensive route into astronomical observing beyond that which is possible with the unaided eye. There are some types of observing, such as estimating the magnitudes of brighter double stars, at which binoculars excel, and there are many deep-sky objects that look significantly better even in a small astronomical binocular than they do in an equivalent-priced telescope. However, if you are considering binoculars as a main instrument, you should take into account that there are aspects of telescope astronomy, such as imaging, which are essentially unavailable to the astronomer who is exclusively a binocular user. Binoculars are limited almost exclusively to visual astronomy, but for sheer enjoyment of the sky, they are unparalleled!

Bibliography

Bradley et al., *Dark-Adapted Pupil Diameter as a Function of Age Measured with the Neuroptics Pupillometer*, Journal of Refractive Surgery, *vol 27(3) March 2011, pp202-7*

Dickinson, T. & and Dyer, A., *The Backyard Astronomer's Guide*, Ontario, Camden House Publishing, 1991, ISBN 0921820119

Fischer, R.E. & Tadic-Galeb, B., *Optical System Design*, New York, McGraw-Hill, 2000, ISBN 0071349162

Gould, J.A., *Journal of the British Astronomical Association*, *vol 80, pp500/1*

Harrington, Philip S., *Touring the Universe through Binoculars*, New York, John Wiley & Sons Inc., 1990, ISBN 0471513377

Salmon, T., http://arapaho.nsuok.edu/~salmonto/VSIII/Lecture11.pdf

Yoder, Paul R., *Mounting Optics in Optical Instruments*, Bellingham, SPIE, 2002, ISBN 0819443328

Chapter 2

Binocular Optics and Mechanics

There are three main parts to a binocular's optical system:

- *Objective lens assembly.* This is the lens assembly at the "big end" of the binocular. Its function is to gather light from the object and to form an image at the image plane.
- *Eyepiece lens assembly.* This is the bit you put to your eyes. Its function is to examine the image at the image plane. The focusing mechanism of the binocular lets you move either the eyepiece assemblies or an intermediate "transfer" lens, so that the eyepieces can focus on the image formed by the objective lenses.
- *Image orientation correction.* In modern binoculars this is usually a prism assembly. Without this, the image would be inverted and laterally reversed, like that in an astronomical telescope. The prisms "undo" this inversion and reversal. In large binoculars, the prism assembly may also enable the eyepieces to be at 45° or 90° to the main optical tube. Binoculars are usually classified by the type of prism assembly they use, e.g., "Porro-prism binocular" or "roof-prism binocular" (Fig. 2.1).

Astronomical observation is exceptionally demanding of optical quality; this applies equally to binoculars as to telescopes, despite the much lower magnification usually used in the former. There are a number of reasons for this demand for higher quality:

- Try this experiment: Make a pinhole of 1 mm diameter or smaller in a piece of paper. Hold this page at a distance where it is just out of focus then, with the book at the same distance from your eye, hold the pinhole up to your eye so that you are now looking through it. Do you see how the page has now come into focus? Astronomy is normally undertaken in the dark, so your eye's pupil is at

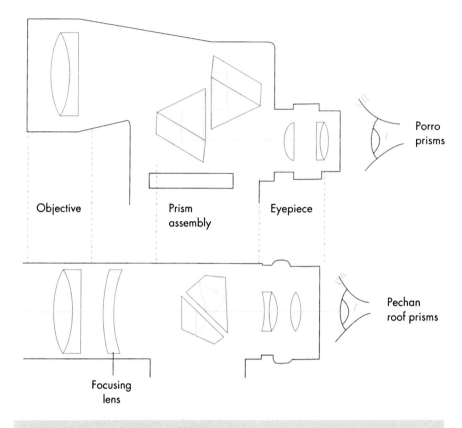

Fig. 2.1 Light-path through prismatic binoculars

its largest. In the daytime, when your eye's pupil is smaller, this smaller aperture can compensate for some optical aberrations in both your eye and the binocular. At night, when the pupil is larger, you do not have this compensation, so any aberrations in the binocular's optics will be much more obtrusive.

- Some visual astronomy involves either objects that are of high contrast with respect to the sky (e.g., double stars). The higher the contrast objects, the higher the demand of optical quality, especially control of chromatic aberration.
- Other visual astronomy involves observing objects of low contrast with respect to the sky (e.g., faint nebulae). Any reduction of contrast in the binocular will make it far more difficult for you to see these objects. All optical aberrations reduce contrast, so these must be kept to a minimum.
- For satisfactory observation of both high- and low-contrast objects, stray light must be minimized. With high-contrast objects, nonimage-forming rays can cause ghost images and reduce contrast if they reach your eye. With low-contrast objects, uncontrolled stray light reduces contrast, rendering the object less visible. Thus, light baffling must be properly designed and implemented, and antireflection coatings of the highest quality should be used on all transmissive surfaces of the optics.

Therefore, unless you are using your binocular only for casual scanning of the sky as a preliminary to using another instrument, it needs to be of the highest optical quality that you can afford. Once you have used a high-quality astronomical binocular, it is very difficult to use one of lesser quality without being dissatisfied, even irritated, by it.

Objective Lens Assemblies

The objective lens consists of two or more lens elements in an achromatic or apochromatic configuration. The achromatic doublet is the commonest lens in "standard" binoculars, but high-quality binoculars, particularly large astronomical binoculars, may have an apochromatic triplet. There may also be additional lenses to correct for other optical aberrations such as spherical aberration (SA), coma, or field curvature. These assemblies containing four or five lenses may be termed "Petzval" lenses, but they are a far cry from the original Petzval lenses, which suffered from a very restricted field of view (about 30°) and a highly curved focal surface. The image "plane" from a simple achromatic or apochromatic lens is actually a curved surface. The purpose of Petzval, and other field-flattening, lenses is to correct the image plane so that it lies on a flat (or, at least, flatter) surface. The binoculars that have these multi-lens designs tend to have coma and field curvature very well controlled. Achromats bring two wavelengths (colors) of light to the same focus. A simple achromatic doublet would have a biconvex element of crown glass in front of a weaker diverging element of flint glass. Modern achromats may use special glasses, such as extra-low dispersion (ED) glass, in order to give better color correction. Apochromats, which bring three wavelengths of light to the same focus, may employ expensive (but brittle) fluorite glass.

Large aperture astronomical binoculars have objectives of relatively small focal ratio, usually as small as f/5, and sometimes less. An achromatic doublet of 100-mm aperture with a focal ratio of f/5 will have significant chromatic aberration, especially off-axis, no matter what glasses are used. This can be particularly obtrusive on bright objects, such as the Moon or the naked-eye planets. Even a fluorite apochromat of this aperture and focal ratio will show off-axis false color on these objects.

Eyepieces

Binocular eyepieces usually consist of three or more lenses in two or more groups. The most common is the venerable Kellner configuration, a design dating from 1849 and which consists of a singlet field lens and a doublet eye lens. Increasingly common are reversed Kellners, a design that was introduced in 1975 by David Rank of the Edmund Scientific Company and used in its RKE eyepieces. The field lens is the doublet and the eye lens is a singlet. The reversed Kellner has the advantages of a slightly wider field (50° as opposed to the 45° of a Kellner), over 50 % more eye relief, and of working better with the short focal ratios that typify binocular objec-

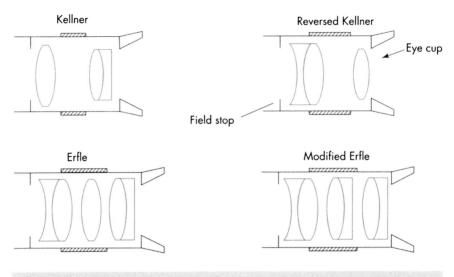

Fig. 2.2 Some common binocular eyepieces

tives. An example of this is the lower-power eyepieces in those 100-mm binoculars, such as the Miyauchi Bj-100B, that have interchangeable eyepieces. Wide-field binoculars usually use modifications of Erfle eyepieces. These consist of five or six elements in three groups. They can have a field of up to about 70°, but eye relief tends to suffer when the field exceeds about 65°. Erfle-type eyepieces have been used extensively in everything from Zeiss Jenoptem and Deltrintem models since 1947 to the current Kunming BA8 models (branded as *Garrett Signature* in the USA and *Helios Apollo* in Europe) (Fig. 2.2).

Prisms

The prisms in binoculars serve primarily to correct the inverted and laterally reversed image that would otherwise result from the objective and eyepiece alone. A secondary effect is that they fold the light path, so that the binocular is shorter than it would otherwise be. For smaller binoculars in particular, this makes them easier to handle. As stated above, binoculars are often classified according to their prism type. For modern binoculars without angled eyepieces, there are two basic types: the Porro prism and the roof prism.

The Porro-prism assembly consists of two isosceles right-angled prisms mounted with their hypotenuses facing each other but with their long axes exactly perpendicular. This latter point is crucial; if they are not exactly at right angles, image rotation (usually referred to as "lean" when it applies to binoculars) will occur. The angle of lean is twice the angle of misalignment and opposite in direction, i.e., a clockwise misalignment of 0.5° will result in an anticlockwise lean of 1.0° (see Fig. 2.3). The light path in Porro prisms is shown in Fig. 2.4. There are four

Fig. 2.3 A rotated prism will cause twice as much rotation in the image

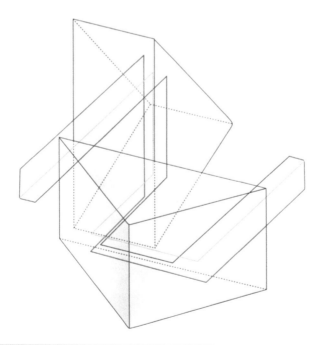

Fig. 2.4 Image inversion and lateral reversal in Porro prism

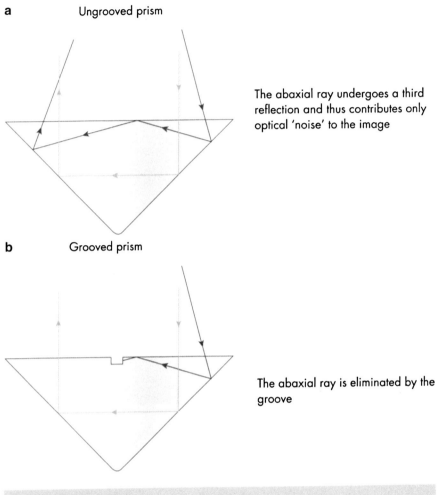

Fig. 2.5 Porro prism groove

reflections, so the result is a right-handed image. The mutually perpendicular orientation of the prism hypotenuses results in one prism erecting the image and the other reverting it.

It is possible, especially when they are used with objectives of low focal ratio, for Porro prisms to reflect rays that are not parallel to the optical axis in such a manner that they are internally reflected off the hypotenuse of the prism (Fig. 2.5a). The ray then emerges from the prism having been reflected a third time and contributes only optical "noise" to the image, thus reducing contrast. This extra reflection can be eliminated by putting a groove across the center of the hypotenuse (Fig. 2.5b). Grooved prisms are a feature of better-quality Porro-prism binoculars.

A development of the Porro prism is the Abbé Erecting System, also known as a Porro type-2 prism, (Figs. 2.6 and 2.7). Its lateral offset is 77 % that of an equivalent

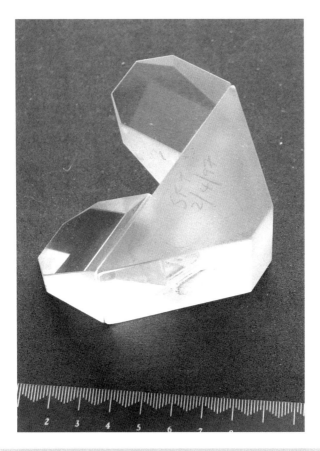

Fig. 2.6 Abbé erecting system, also known as a Porro type-2 prism

Porro-prism assembly,[1] and for this reason, it is most frequently encountered in larger binoculars which would otherwise have to have their objective lenses more widely spaced to allow the eyepieces to have a usable range of interpupillary distance. For medium-aperture binoculars, it is more common in older instruments, particularly military binoculars from the early and mid-twentieth century. Abbé Erecting Systems are usually identifiable by the cylindrical prism housing, although the reverse is not true, i.e., this feature is not diagnostic of the presence of the Abbé system.

Another consideration is the glass used for the prism. Normal borosilicate crown (BK7—the *BK* is from the German *Borkron*) glass has a lower refractive index than the barium crown (BaK4—the *BaK* is from the German *Baritleichtkron*) glass that is used in better binoculars. A higher refractive index results in a smaller critical

---

[1]Yoder 2002

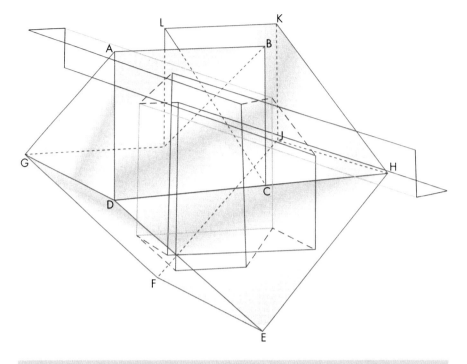

Fig. 2.7 Light path in Abbé erecting system (aka Porro type-2)

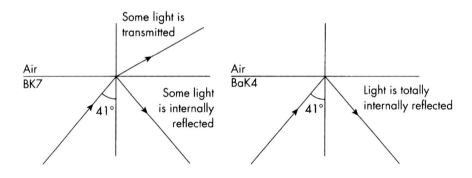

Fig. 2.8 Bk7 and BaK4 glass. At angles close to the critical angle of Bak4 glass, some light will be lost due to transmission in BK7 glass

angle, 39.6° in BaK4 as compared to 41.2° in BK7, so there is less light likely to be lost because of non-total internal reflection in the prisms (Fig. 2.8). The difference is more noticeable in wide-angle binoculars whose objective lenses have a focal ratio of f/5 or less. The non-total internal reflection of the peripheral rays of light cone from the objective results in vignetting of the image. This effect can easily be

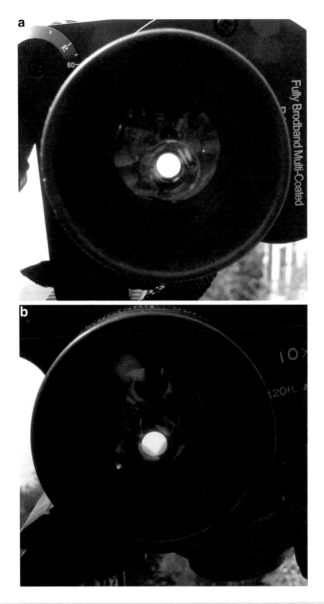

Fig. 2.9 The effect of prism glass on the exit pupil. (**a**) BAK4 prisms. (**b**) BK7 prisms

seen by holding the binocular up to a light sky or other light surface and examining the exit pupil. The exit pupil of a binocular with BaK4 prisms will be perfectly round, while that of a binocular with BK7 prisms will have telltale blue-gray segments around it (Fig. 2.9). (Note: Fig. 2.9b was taken from a slight angle in order to show the nature of the vignette segments. Viewed from directly behind the exit pupil, there is a square central region with vignette segments on four sides.)

Glass Type	Refractive Index	Critical Angle	Dispersion
Schott BaK4	1.5688	39.6°	-0.0523 µm^{-1}
Chinese BaK4	1.5525	40.1°	-0.0452 µm^{-1}
Schott BK7	1.5168	41.2°	-0.0418 µm^{-1}

Fig. 2.10 Specifications of some common prism glass

However, BaK4 glass has a lower Abbé number than Bk7 glass. This means that any rays that are not normal (perpendicular) to the prism when they enter or exit it will be dispersed more by BaK4 glass than by BK7 glass. At the magnification in most binoculars, you are unlikely to be able to detect this in use, but it is one of the reasons that BK7 prisms may be a preferable prism material for specialist high-power binoculars.

It is important to recognize that the prism glass is but one of the many considerations that affect image quality. There are excellent older binoculars that use BK7 glass for the prisms and which give a better image quality than many of the modern budget offerings that have "BaK4" printed on their cover plates. BK7 is also the glass of choice for binoviewer prisms, owing to its lower dispersion and the lack of need to accommodate wide-angle use.

Bak4 is a glass designation used by Schott AG, an old and respected German manufacturer of optical glass. Although there are international standards for optical glass designation, BaK4 isn't one of them. Anyone can apply it to any glass. The international standard designation for Schott BaK4 is 569561. The first three digits tell you its refractive index (1.569) and the last three tell you its Abbé number (56.1), which indicates how much it will disperse light into its component colors; the higher the Abbé number, the less the dispersion. However, I don't see customers being willing to learn and compare international standard designation codes: "Bak4" trips off the tongue so much more easily.

This is what you should know: the "BaK4" glass used for the prisms of Chinese binoculars is not the same as Schott BaK4. In fact, it's not even barium crown, which is what BaK stands for! It is a phosphate crown glass with a lower refractive index and dispersion than Schott BaK4 (but higher than BK7). It also potentially has a higher "bubble count" (Fig. 2.10).

In practice, this may not be all bad. Unless you have very wide-angle binoculars, you are unlikely to notice the effect of the lower refractive index, and the lower dispersion than "real" BaK4 means that there may be less dispersion in the image (not that you are likely to be able to see it). The potentially higher bubble count means there may be more light scatter inside the prism; I've not been able to detect it in use.

The roof prism is shown in Fig. 2.11. It is a combination of a semi-pentaprism (45° deviation prism) (Fig. 2.12) and a Schmidt roof prism (Fig. 2.13). The combination is a compact inversion and reversion prism that results in an almost "straight-through" light path. The consequence is a very compact binocular. There is, of

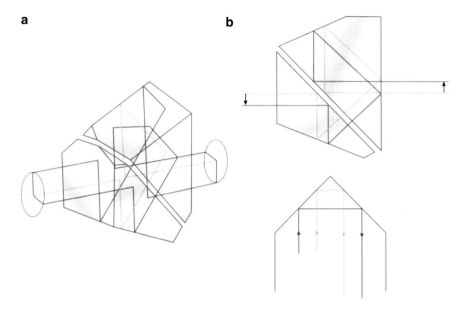

Fig. 2.11 Image reversal in Pechan roof prism

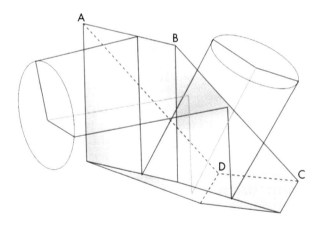

Fig. 2.12 Semi-pentaprism (45° deviation prism)

course, a limit to the aperture of roof-prism binoculars that is imposed by the "straight-through" light path because, the centers of the objectives cannot be separated by more than the observer's interpupillary distance (IPD).

Although the roof-prism configuration is physically smaller and thus uses less material in its construction, it tends to be significantly more expensive than a Porro-

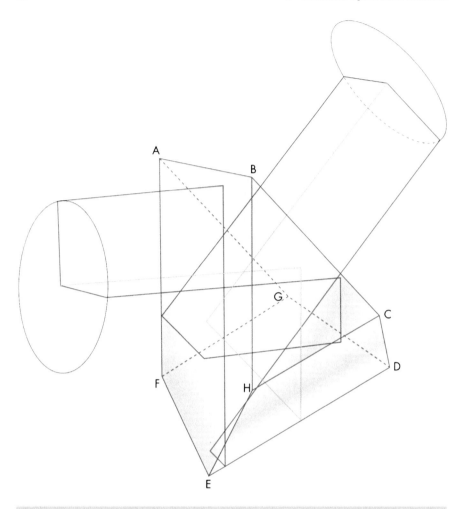

Fig. 2.13 Schmidt roof prism. The image is inverted and reverted. The axis is deviated by 45°

prism binocular of equivalent optical quality. This is because the prism system, particularly the roof itself, must be made to a much higher tolerance (2 arcsec for the roof) than is acceptable for Porro prisms (10 arcmin), i.e., 300 times as precise! Any thickness or irregularity in the ridge of the roof will result in visible flares, particularly from bright high-contrast objects, i.e., many astronomical targets. Additionally, a result of the wave nature of light is that interference can occur when a bundle (aka pencil) of rays is separated and recombined, as happens with a roof prism. The consequence is a reduction in contrast. This can be ameliorated by the application of a "phase coating" to the faces of the roof. Binoculars with phase coatings usually have "PC" as part of their designation (see Appendix 6).

As you will see from Fig. 2.11, the light in a Schmidt-Pechan roof prism undergoes six reflections (as opposed to four in a Porro-prism binocular). This results in a "right-handed" image. A consequence of the extra reflection and the extra focusing lens (as compared to Porro prisms) is more light loss. In order to achieve a similar quality of image, better antireflective coatings need to be used. The Abbe-König prisms used in some better-quality roof-prism binoculars have only four reflections, and the prism thus transmits about 2 % more light than the Schmidt-Pechan.

The demand for better quality of the optical elements and their coatings in roof-prism binoculars means that they will inevitably be more expensive than Porro-prism binoculars of equivalent optical quality. They do, however, offer three distinct advantages:

- They are more compact. This makes them slightly easier to pack and carry; some people (I am one) find the smaller size easier and more comfortable to hold as a consequence of the different ergonomics.
- They are usually slightly lighter. This makes them easier to carry and generally less tiring to hold.
- They are easier to waterproof as a consequence of the internal focusing. Although one does not normally do astronomy in the rain (the possible exception being the nocturnal equivalent of a "monkeys' wedding"), nitrogen-filled waterproof binoculars are immune to internal condensation in damp/dewy conditions and will not suffer from possible water penetration when used for other purposes such as bird-watching or racing.

It is a matter of personal judgement whether these advantages warrant the extra expense. I find that, on account of their relative lightness and compactness, I observe with my 10×42 roof prisms far more than I do with my 10×50 Porro prisms.

There is a common misconception that roof-prism binoculars are "birding binoculars" and that Porro-prism binoculars are inherently better for astronomy. Whereas roof-prism binoculars are advantageous for birding (lighter, easier to waterproof) and Porro-prism binoculars generally offer equivalent optical quality at a lower price and are not aperture limited because of the design, both can be used for either activity, where the one with the better optical quality will generally perform better. The best handheld binocular I have used for astronomy is a Swarovski EL 10×50 (roof prism): it was light and well balanced, very bright, and had no noticeable aberrations.

An increasing number of astronomical binoculars have $45°$ or $90°$ eyepieces. There are a wide variety of prism combinations that will achieve this, such as a Porro type 2 with a semi-pentaprism for $45°$ or with a pentaprism for $90°$. Another $45°$ system uses a Schmidt roof with a rhomboid.

Binoviewers use a combination of a beam splitter and a pair of rhomboidal prisms (Figs. 2.14 and 2.15). The beam splitter divides the light equally into two mutually perpendicular optical paths. A rhomboidal prism merely displaces the axis of the light path without either inverting or reverting it. In some binoviewers,

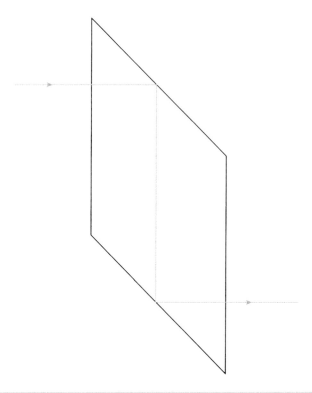

Fig. 2.14 Rhomboid prism. This prism displaces the axis

pairs of mirrors perform the same function. Cylindrical light tubes may be used to ensure that the optical path length is identical on both sides. Interpupillary distance is adjusted by hinging the device along the axis of the light path from the objective lens or primary mirror.

Some observers use image-stabilized binoculars. Image stabilization was first introduced for camera lenses and for military surveillance; the technology was later transferred to astronomical binoculars.

The system of image stabilization that has been most successful for astronomical purposes is that developed by Canon Inc. It employs what Canon calls a Vari-Angle Prism (Fig. 2.16) which consists of two circular glass plates that are joined at their edges by a bellows of a specially developed flexible film. The intervening space is filled with a silicon-based oil of very high refractive index. Microelectronic circuitry senses vibration and actuates the Vari-Angle Prisms so as to compensate for the change in orientation of the binoculars.

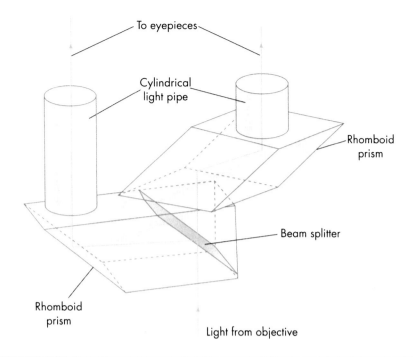

Fig. 2.15 The principle of binoviewer

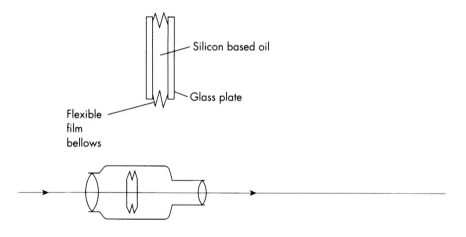

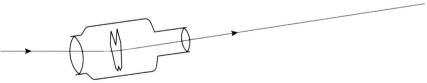

Fig. 2.16 Image stabilization with Canon's vari-angle prism

Coatings

The most important coatings in binoculars are the antireflective coatings on the surfaces of the optical components. An uncoated glass-to-air surface will reflect about 4 % of the light that is perpendicular to it ("normal incidence") and even more of the light that is oblique to it. By using interference coatings, this can be reduced to better than 0.15 % over a very wide range of the optical spectrum. The coatings are usually optimized for a particular wavelength of light, usually in the range 510–550 nm, which is the yellow-green part of the spectrum where the human eye is most sensitive to light. If the coating is optimized, intentionally or otherwise, for another part of the spectrum, the image will have a color cast. An extreme example of this is the "ruby" coatings found on some very low-quality binoculars, where the coating serves to remove light from one end of the visible spectrum in an attempt to conceal the poor color correction that is inherent in a cheap and inadequate optical design. A single coating of a quarter, the wavelength of light will reflect a small proportion of the incident light. The glass behind it will reflect another small proportion. The path length of the wave reflected off the glass is half (2×¼) a wavelength greater; the two reflected waves mutually interfere destructively, eliminating the reflection for that particular wavelength (Figs. 2.17 and 2.18). At wavelengths significantly distant from the wavelength for which the coating is optimized, interference may be constructive, resulting in more reflected energy than would have occurred in uncoated glass. Additional layers of half- and quarter-wave thickness can reduce reflections at other wavelength; this is

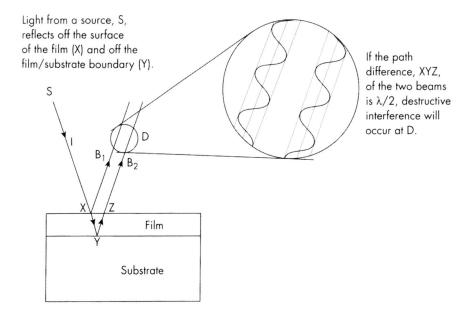

Fig. 2.17 Single layer flim

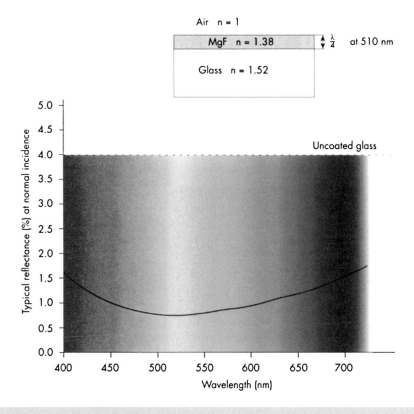

Fig. 2.18 Coated optics: single layer coating

"multicoating" (Fig. 2.19) and "broadband multicoating" (Fig. 2.20). Each additional layer of coating has a progressively lesser effect on improving light transmission. Coating is an expensive process, so there are a number of coatings that become uneconomical. It is rare to find more than seven layers on any surface in commercial binoculars.

Binocular coatings are qualitatively described as "coated," "fully multicoated," etc. There is no universally agreed meaning to these designations, but they are commonly held to have the following meanings:

- **Coated**: At least one glass-to-air surface (usually the outer surface of the objective) has a single layer of antireflective coating, usually MgF_2; other surfaces are uncoated.
- **Fully Coated**: All glass-to-air surfaces of the lenses (but not the prism hypotenuses) have a layer of antireflective coating.
- **Multicoated**: At least one glass-to-air surface (usually the outer surface of the objective) has two or more layers of antireflective coating. The other surfaces may be single-layer coated or not coated at all.
- **Fully Multicoated**: All glass-to-air surfaces of the lenses (but possibly not the prism hypotenuses) have two or more layers of antireflective coating.

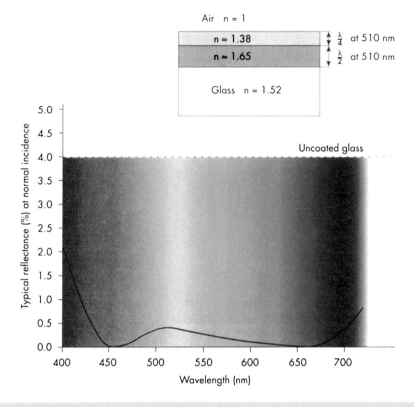

Fig. 2.19 Multi-coated optics: double layer coating

More recently, some binocular coatings have been described as "broadband." Again, there is no industry-wide standard—it can mean anything from three layers upward. Some manufacturers are more forthcoming as to the precise nature of their coatings. For example, Kunming Optical, the manufacturer of the popular *Garrett Optical* and *Oberwerk* binoculars in the USA (branded as *Strathspey* and *Helios Apollo* in the UK, *Teleskop-Service* in Germany), provides the following information about its coatings[2]:

- **Level I**: (Equivalent to *fully coated*) Single layer of MgF_2 coating on 16 glass-to-air surfaces— four for two objectives, 12 (6 per side) for the three optical elements in each eyepiece. The prisms are not coated.
- **Level II**: (Equivalent to a blend of *multicoated* and *fully multicoated*) Broadband multicoatings of 5–7 layers on the four glass-to-air surfaces of the two objectives and the four surfaces of the eye lenses of the two eyepieces. Single-layer MgF_2 coating on all other glass-to-air surfaces, including the hypotenuses of the prisms.

---

[2]Kunming Optical Instrument Co. Ltd.

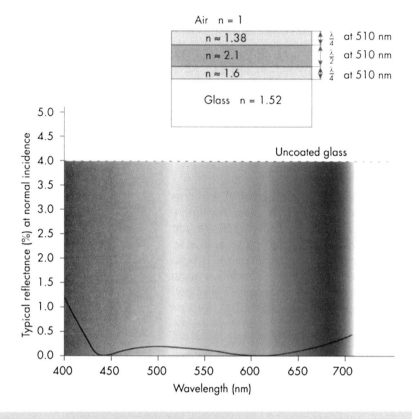

Fig. 2.20 Broadband multi-coated optics: triple layer coating

- **Level III**: Broadband multicoatings on all the surfaces except the prism hypotenuses, on which there are single-layer MgF_2 coatings.
- **Level IV**: Broadband multicoatings on all the surfaces including the prism hypotenuses.

The effect of various coatings can be seen in the reflections of sunlight from objective lenses in Fig. 2.21.

One of the criteria that is often offered, by well-meaning people, as an important consideration in binocular choice is that it should be "fully multicoated." Coatings are only effective if they are properly designed and applied. In budget binoculars, they are often unevenly applied (sometimes giving a "patchy" appearance to the lens surface if the unevenness is extreme), so are less effective. The quality control of these items is also usually extremely cursory in nature, so "fully multicoated" has a reduced value, and there are other criteria, such as control of aberrations and stray light, that are much more important in this class of binocular.

The two 70-mm binoculars in Fig. 2.22 were made in the same factory and ostensibly have the same broadband fully multicoated optics. I photographed

Fig. 2.21 Optical coatings. Clockwise from top left: single-layer coated, broadband multi-coated, multicoated, uncoated

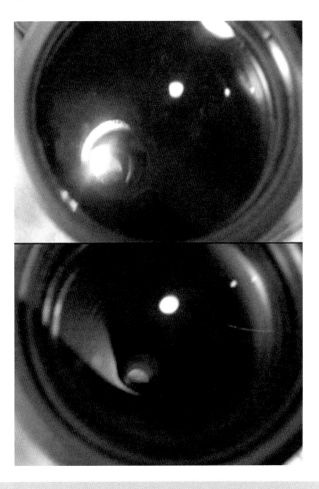

Fig. 2.22 "Fully multicoated" does not always mean the same thing

them under similar conditions. One is three times the cost of the other. Guess which is which.

Aberrations

Aberrations are errors in an optical system. There are six optical aberrations which may affect the image produced by a telescope. Some affect the quality of the image; others affect its position. They are:

- Chromatic aberration: error of quality
- Spherical aberration: error of quality
- Coma: error of quality
- Astigmatism: error of quality
- Field curvature: error of position
- Distortion: error of position

Chromatic aberration is an error of refractive systems and is therefore of consideration for all binoculars. Because any light which does not impinge normally on a refractive surface will be dispersed, single converging lenses will bring different wavelengths (colors) of light to different foci, with the red end of the optical spectrum being most distant from the lens. This is *longitudinal* (or *axial*) chromatic aberration. It usually manifests itself as a colored halo, which changes in color from purplish at best focus to greenish outside focus (known as the "apple and plum" effect), around bright objects. *Lateral* chromatic aberration manifests as different wavelengths of light forming different sized images. It usually manifests itself as *color fringing* on off-axis objects. The term *color fringing* is descriptive of the visual effect of its presence.

Visible chromatic aberration can exist in objective lenses and eyepieces. Chromatic aberration can be reduced, but not eliminated, by using multiple lens elements of different refractive indices and dispersive powers. An achromatic lens has two elements and brings two colors to the same focus (Fig. 2.23).

The choice of glass and lens design will determine not only which colors are brought to the same focus but also the distance over which the secondary spectrum is focused. An apochromatic lens uses three elements and will bring three colors to the same focus. Using nonexotic glass, each additional lens will reduce chromatic aberration by about 80 %. Hence, an achromatic doublet can be expected to have approximately 20 % of the chromatic aberration of a singlet lens. An apochromatic triplet will reduce it to 20 % of the achromat's 20 %, i.e., approximately 4 % of the chromatic aberration of the equivalent singlet. The use of exotic glasses such as fluorite or ED will reduce it even further, to the extent that, say, an ED doublet may have less than 10 % of the chromatic aberration of the equivalent singlet. Such a combination is often termed a "semi-apochromatic."

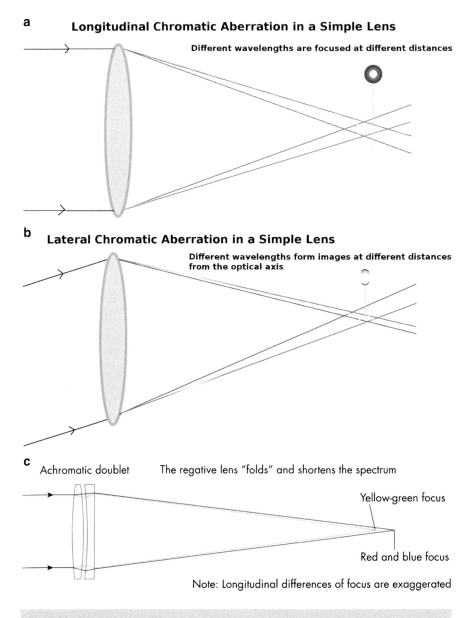

Fig. 2.23 Chromatic aberration

Spherical aberration is an error of spherical refractive and reflective surfaces which results in peripheral rays of light being brought to different foci to those near the axis (Fig. 2.24).

If the peripheral rays are brought to a closer focus than the near-axial rays, the system is *undercorrected*. If they are brought to a more distant focus, the system is

Aberrations

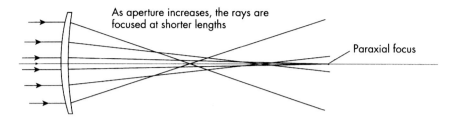

Fig. 2.24 Spherical aberration in a converging lens

overcorrected. Spherical mirrors and converging lenses are undercorrected and diverging lenses are overcorrected.

In compound lenses, spherical aberration can be suppressed in the design of the lens, by using several lenses of minimal curvature as a substitute for one of considerable curvature, by choosing appropriate curvatures for the converging and diverging elements, or as a combination of both. In Newtonian mirrors, such as are used in most reflecting binocular telescopes, the spherical aberration is corrected by progressively deepening the central part of the mirror so that all regions focus paraxial rays to the same point. The shape of the surface is then a *paraboloid*, that is, the surface that results from a parabola being rotated about is axis.

There are other manifestations of spherical aberration, the most common of which is zonal aberration, in which different zones of the objective lens or primary mirror have different focal lengths.

Spherical aberration increases as a direct cubic function of increase in aperture and is independent of field angle.

Coma can be considered to be a sort of a lopsided spherical aberration. If an objective lens is corrected for paraxial rays, then any abaxial ray cannot be an axis of revolution for the lens surface and different parts of the incident beam of which that ray is a part will focus at different distances from the lens. The further off-axis the object, the greater the effect will be. The resulting image of a star tends to flare away from the optical axis of the telescope, having the appearance of a comet, from which the aberration gets its name. In objective lenses, coma can be reduced or eliminated by having the coma of one element counteracted by the coma of another. It is usually particularly noticeable in ultrawide-angle binoculars.

Coma often occurs in combination with astigmatism (see below) (Fig. 2.25).

Coma increases as a direct square (quadratic) function of aperture increase and as a linear function of increase in field angle.

Astigmatism results from a different focal length for rays in one plane as compared to the focal length of rays in a different plane. A cylindrical lens, for example, will exhibit astigmatism because the curvature of the refracting surface differs for the rays in each plane and the image of a point source will be a line (Fig. 2.26).

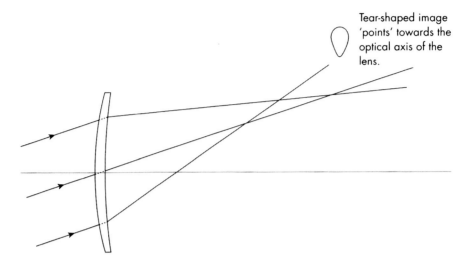

Fig. 2.25 Coma in a converging lens

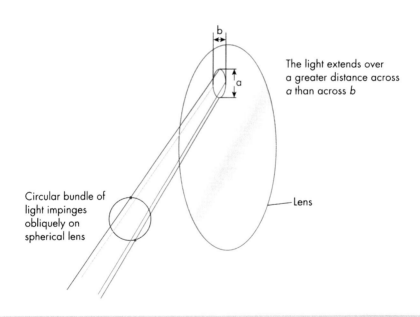

Fig. 2.26 Astigmatism

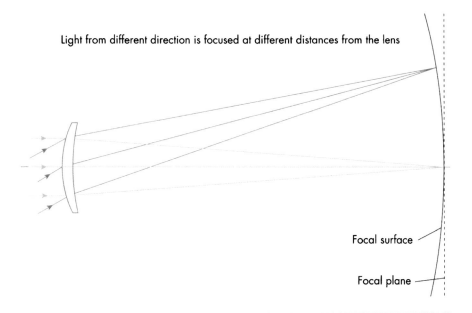

Fig. 2.27 Field curvature. *Amend text at top of image to read: "Light from different directions is focused at different horizontal distances from the lens"*

Astigmatism will therefore result from any optical element with a surface which is not a figure of revolution.

It can also occur in surfaces which *are* figures of revolution. Consider two mutually perpendicular diameters across a beam of light impinging obliquely upon a lens surface. The curvature of the lens under one diameter differs from that under the other, and so astigmatism will occur. Such astigmatism can be corrected by an additional optical element which introduces equal and opposite astigmatism. Astigmatism is not normally a problem in binoculars, which are primarily used for visual work, unless they have very wide fields.

Astigmatism rarely occurs alone and is usually combined with coma; the combined effect is that star images, especially near the periphery of the field of view, appear as "seagulls," i.e., there is a curved "wing" apparent to each side of the center of the star image.

Astigmatism increases linearly with increase in aperture and as a direct square (quadratic) function of increase in field angle.

Field Curvature. No single optical surface will produce a flat image—the image is focused on a surface which is a sphere which is tangential to the focal plane at its intersection with the optical axis (Fig. 2.27).

Field curvature, which manifests as the inability to focus the periphery of the image at the same time as the center is focused, is particularly noticeable when it is present in wide-field binoculars. It can be corrected in the design of the lenses. In particular, if a negative lens can be placed close to the image plane, it will flatten the field.

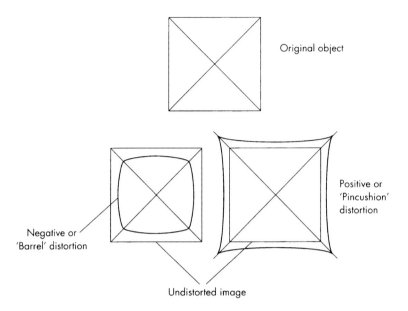

Fig. 2.28 Distortion

Field curvature increases linearly with increase in aperture and as a direct square (quadratic) function of increase in field angle.

Distortion is an aberration by which a square object gives an image with either convex lines (*negative* or *barrel distortion*) or concave lines (*positive* or *pincushion distortion*). It is the only aberration that does not produce blurring of the image (Fig. 2.28).

It results from differential magnification at different distances from the optical axis. It almost always originates in the eyepiece, so any correction should be inherent in eyepiece design. A small amount of pincushion distortion can be desirable because it attenuates the "rolling ball" effect that results in an undistorted field of view (see Chap. 3). This effect, which strictly speaking appears as a "rolling cylinder" as the binocular is panned across the sky, with the axis of the cylinder perpendicular to the direction of panning, can be disorientating and unpleasant, to the extent that it causes nausea in some observers.

Distortion is unaffected by aperture and increases as a direct function of the cube of field angle.

Aperture Stops and Vignetting

Vignetting is the loss of light, usually around the periphery of an image, as a consequence of an incomplete bundle passing through the optical system. A vignetted image appears dimmer around the periphery.

Most binoculars suffer from some degree of vignetting. The exception is some binoculars designed specifically for astronomical use and whose construction is based on astronomical refracting telescopes which themselves give unvignetted images. An example of this is the 22×60 Takahashi Astronomer, which used two Takahashi FS60 optical tubes

In some, it can be so severe that no part of the image is illuminated by the complete aperture. In normal daylight use, we do not notice vignetting unless it is exceptionally severe; 30 % is common and 50 % is sometimes deemed acceptable in wide-angle systems. This is because, at any given time, only a tiny region of the image can be examined by the fovea and it is therefore only this region that needs to be fully illuminated. As long as the fall-off of illumination towards the periphery is smooth, it will not normally be noticed.

Binocular astronomers who, like other astronomers, echo the call for "More light!" sometimes wonder why vignetting is allowed to occur at all. To understand this, we must first understand the role of the aperture stop. An aperture stop crops the light cone and eliminates the most peripheral rays. These peripheral rays have the highest angles of incidence on the optical surfaces and undergo the most refractive bending. For these reasons, they also carry with them the greatest amount of aberration. If they are permitted to pass through to your eye, they will add to the degradation of the image. Part of the process of good optical design is to assess how much of the peripheral light needs to be excluded.

If bundles of rays from all parts of the field of view fill the aperture stop, then there is no vignetting. On the other hand, if some other mechanical or optical component impedes some of this light, vignetting will occur. An unvignetted binocular requires larger optical apertures all the way through the optical system when compared to one in which vignetting does not occur. This in turn requires larger optical components (such as prisms or focusing lenses). Larger components are not only more expensive but also heavier. Heavier components require heavier and more robust mountings. These in turn add to the expense of the binocular. The overall result is a heavier, more expensive, binocular. In short, vignetted systems are usually smaller and lighter and produce better images in comparison to the equivalent unvignetted optical system. Somewhere in the design process, a decision is made as to where an acceptable trade-off lies. The more discerning observer may well be prepared to accept a more expensive instrument, but the general user will almost certainly not want to pay considerably more for a hardly noticeable increase in light throughput at the periphery. Even the discerning observer may balk at an increase in weight if the binoculars are intended to be handheld.

Focusing Mechanisms

There are three different types of focusing mechanism commonly found on binoculars:

Center Focus (Porro Prism)

The eyepieces are connected to a threaded rod in the central hinge. An internally threaded knurled wheel or cylinder causes the rod to move, thus moving the eyepieces. The right-hand eyepiece is usually independently focusable (Fig. 2.29a) in

Fig. 2.29 Right eyepiece diopter adjustment. (**a**) Porro prism. (**b**) Roof prism

order that differences in focus of the observer's eyes can be accommodated; this facility is often called a "diopter adjustment." The advantage is that the eyepieces can be focused simultaneously, which is a consideration for general terrestrial use, but not for astronomy. The disadvantages are that there is almost always some rocking of the bridge, which leads to difficulty in achieving and maintaining focus; the focusing system is difficult to seal, so dirt can enter; and the optical tubes are extremely difficult to waterproof, resulting in increased likelihood of internal condensation (Fig. 2.30).

Center Focus (Roof Prism)

Like the Porro-prism center focus system, there is an external focus wheel and an independent helical focuser (diopter adjustment) for the right eyepiece (Fig. 2.29b); the similarity ends there. The mechanism is internal and focusing is achieved by changing the position of a focusing lens between the objective lens and the prism assembly (Fig. 2.1). It has the dual advantages of permitting simultaneous focusing of both eyepieces and allowing relatively simple dust- and waterproofing. The disadvantage is that there is an extra optical element that must be accurately made, which absorbs a tiny amount of light, and whose movement during focusing alters the field of view slightly (Fig. 2.31).

Fig. 2.30 The bridge rocks on this center-focus Porro-prism binocular

Fig. 2.31 Roof-prism center focus

Independent Focus

The eyepieces each have a helical focuser. This is much more robust than a center focus system and is easier to make dirt- and waterproof. The best-quality astronomical (and marine and military) binoculars have independent focusing. The disadvantage is that the eyepieces cannot be focused simultaneously, but this is not an issue for astronomical observation, where refocusing is not necessary once good focus has been attained (Fig. 2.32).

Collimation

Fig. 2.32 Independent focus—ideal for astronomy

Collimation

Not only must the optical elements of each optical tube be collimated, but the optical axes of both tubes must be aligned. They must not only be aligned to each other but also to the hinge or other axis about which interpupillary distance is adjusted. If this latter criterion is not met, the result is a phenomenon called *conditional alignment* in which the two optical axes are only aligned at the interpupillary distance that was set during collimation and will get progressively out of alignment for other interpupillary distances. This may be acceptable if only one person is to use the binocular, but should never be so bad that the exit pupils take on a "cat's eye," as opposed to circular, appearance.

The permitted divergence of the optical axes from true parallelism is determined by the ability of the eyes to accommodate divergence and by the magnification of the binoculars. If these limits are exceeded, either it will not be possible to merge

Table 2.1 Collimation tolerances

Magnification	Step (arcmin)	Convergence (arcmin)	Divergence (arcmin)
×7	2.0	6.5	3.0
×10	1.5	4.5	2.0
×15	1.0	3.0	1.5
×20	0.75	2.25	1.0
×30	0.5	1.5	0.67
×40	0.38	1.13	0.5

the images from each optical tube or, if they can be merged, eyestrain and its attendant fatigue and/or headache results. Acceptable tolerances in the apparent field of view are as follows:

- Vertical misalignment (step, dipvergence): 15 arcmin
- Horizontal convergence[3]: 45 arcmin
- Horizontal divergence: 20 arcmin

To ascertain the real tolerances, you need to divide these by the magnification, to obtain the collimation tolerances listed in Table 2.1.

There are two ways in which the optical axes of the binoculars can be aligned. In almost all binoculars, the objective lenses are mounted in eccentric rings. These can be adjusted to move the optical axis in relation to the body of the binocular. In many other binoculars, the prisms are adjustable, either by grub screws (set screws) that are accessible from the outside or by being housed in a cluster whose adjustment screws are accessible by removing the cover plate on the prism housing. (See Chap. 5 for advice on how to collimate a binocular.) Collimation by eccentric rings on the objectives is preferable, because tilting the prisms will result in the introduction of more astigmatism.

Bibliography

Fischer, R.E. & Tadic-Galeb, B., *Optical System Design*, New York, McGraw-Hill, 2000, ISBN 0071349162

Kunming Optical Instrument Co. Ltd., http://www.binocularschina.com/

Lombry, T., http://www.astrosurf.com/lombry/reports-coating.htm

---

[3]There are different conventions for the use of "convergence" (and "divergence"), depending on whether the optical axes of the binocular are converging or the optical axes of the eyes are converging (to accommodate the diverging optical axes of the binocular). The usage here is the latter. It is simple to tell which convention is being used: the greater value is for converging eyes (diverging binoculars).

The Naval Education and Training Program Development Centre, ***Basic Optics and Optical Instruments***, New York, Dover, 1997, ISBN 0-486-2291-8.

Pedrotti, F.L. & Pedrotti, L.S., ***Introduction to Optics,*** Englewood Cliffs, Prentice-Hall Inc., 1993, ISBN 0-13-016973-0

Tonkin, Stephen F., ***AstroFAQs***, London, Springer-Verlag, 2000, ISBN 1-85233-272-7

Yoder, Paul R., ***Mounting Optics in Optical Instruments***, Bellingham, SPIE, 2002, ISBN 0819443328

Chapter 3

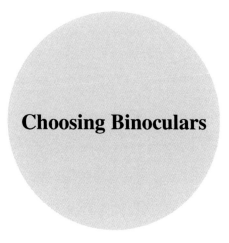

Choosing Binoculars

There is no "ideal" binocular for astronomy; the individual choice is therefore determined by the *reason* for choosing binoculars, the *purpose* to which the binoculars will be put, and the *budget*.

Deciding What You Need

Standard advice given, in good faith, on the Internet and in magazines is usually along the lines of "if you get fully multicoated (FMC) and BaK4 prisms, you won't go far wrong." This is very much not the case—there are at least 20 relevant things that "fully multicoated" (FMC) and "BaK4 prisms" tells you exactly nothing about. These include, in no particular order:

- Quality of internal light baffling
- Type and quality of eyepieces
- Field curvature
- Spherical aberration
- Crispness of focus
- Edge distortion
- Amount of vignetting
- Size of fully illuminated field of view
- Chromatic aberration
- Mechanical build quality
- Smoothness of focus
- Manufacturer's quality control

- With respect to coatings:
  - Evenness of application
  - Whether they are the correct thickness
  - Whether there are seven layers on all glass-air surfaces, including prism hypotenuses, or whether it's just two layers on the glass-air surfaces of the lenses
- With respect to the prisms:
  - Whether it's Schott BaK4 or Chinese BaK4 glass.
  - If the prisms are undersized. If they are, they will cut out some light.
  - The precision with which the flat surfaces of the prism have been polished.
  - Whether the prism hypotenuses are grooved. Grooved prisms reduce spurious reflections.
  - Whether the prism sides are blackened. Prevents nonimage-forming light entering the prism.
  - Whether the reflective surfaces of the prisms are shielded. Prevents nonimage-forming light entering the prism.
  - How the prisms are secured into their housings. There's an enormous difference between glue and a properly constructed prism cage.

I would far rather have a binocular with properly applied single coatings and full-sized, precisely polished BK7 prisms with blackened sides, grooved hypotenuses, and shielded reflective surfaces that are held in a proper cage than one with shoddily applied multilayer coatings and undersized "naked" ungrooved BaK4 prisms that have been made with minimal quality control and are glued into the housing and held in place with a spring clip.

Binocular Specifications

Binoculars are specified by a series of numbers and letters, e.g., 15×70 BIF. GA.WA. The numbers tell you the size of the binocular and the letters give additional information. The first number is the angular magnification; the second is the aperture of the objective lens in millimeters. The example above therefore has a magnification ("power") of 16, an aperture of 70 mm, a body of "Bausch & Lomb" (aka "American") construction (B), individually focusing eyepieces (IF), rubber armor (GA), and wide-angle eyepieces (WA). There is a complete list of designation letters in Appendix 6.

The numbers in the binocular specifications give rise to a variety of binocular ratings that are sometimes quoted by manufacturers and vendors. The most common are:

- **Relative Brightness**. This is the square of the diameter of the exit pupil. The exit pupil diameter is calculated by dividing the aperture by the magnification (power). For example, for 10×50, the exit pupil is 5 mm and the relative brightness is $(50/10)^2 = 5^2 = 25$. However, the calculation for a 20×100 binocular, through which a great deal more can be seen, gives exactly the same relative brightness: $(100/20)^2 = 5^2 = 25$, so this is an inadequate rating to use for astronomical binoculars. Incidentally, this is also the relative brightness that is calculated for the Mark-I eyeball (1×5) of a human being approximately 60 years old! Whereas it does give information about the surface brightness of extended objects, it says little about the overall performance; I have no doubt that I see far more in my 10×50 binoculars than I do with the naked eye and that I see significantly more in my 20×100 binocular.

- **Twilight Index** or **Twilight Performance Factor**. This was used by *Carl Zeiss International* as an indication of the *distances* at which comparable detail would be seen in different binoculars. It is calculated by finding the square root of the product of the magnification and aperture. For the two binoculars above, the calculations are:

$$\sqrt{(10 \times 50)} = \sqrt{(500)} = 22.36$$

$$\sqrt{(20 \times 100)} = \sqrt{(2,000)} = 44.72$$

In this instance, the larger instrument has an index that is double that of the smaller instrument. In other words, if the smaller binocular is used to observe a target object at a given distance from the observer, the same amount of detail will be visible at double the distance in the larger instrument. This is not really applicable to visual observational astronomy where we are usually not concerned with the relative distances of objects and consider them to be effectively at the same distance.

- **Visibility Factor**. This is due to Roy Bishop[1] and is evaluated simply by multiplying the magnification by the aperture in millimeters. For our two binoculars above, we obtain:

$$10 \times 50 = 500$$

$$20 \times 100 = 2,000$$

The larger instrument has a visibility factor four times greater than the smaller one. Bishop justifies this by stating: "in the larger instrument stars will be four times brighter and extended images will have four times the area from which the eyes can glean information, with luminances being the same."[2] While this is objectively correct, I am not convinced that it reflects the subjective experience of observing through both instruments.

- **Astro Index**. This is due to Alan Adler[3] and is evaluated as the product of the magnification and the square root of the aperture in millimeters. For our two binoculars above, we obtain:

$$10 \times \sqrt{50} = 10 \times 7.1 = 71$$

---

[1] Bishop, 2002, p. 50
[2] Bishop, op cit, p. 51
[3] Adler, 2002

$$20 \times \sqrt{100} = 20 \times 10 = 200$$

This gives the larger instrument an Astro Index of 2.8× the smaller one; this is certainly closer to the experience of observing through them.

Others[4] have tried to expand on these by including the effects of coatings, baffles, and other aspects of individual quality, but, although these may be more precise, they incorporate a certain amount of empirical experimental data and are not as valuable for a quick evaluation, prior to purchase, of likely performance.

What Size?

There is a baffling array of binocular sizes, many of which are potentially useful for astronomy. Although you will see more through even a 20-mm binocular than you will with your unaided eye, these very small binoculars cannot really be regarded as "good" for astronomy. I tend to class binoculars for astronomy into four different categories (based on size and magnification), each of which is approximately a magnitude brighter than the previous one. Roughly speaking, these are:

Ultra-portable: Above 35 mm but less than 50-mm aperture with a magnification of 7× or 8×. These are lightweight, have wide fields of view, typically in excess of 7°, and are easy to handhold. Except under really dark skies, they are unlikely to be usable as a primary observing instrument, but are excellent for quick scans of the sky to ascertain sky conditions, or for orientating yourself with respect to fainter star fields as a preliminary to using, say, a telescope. Typical examples are 7×35 and 8×40. Note that Fr Lucian Kemble discovered his eponymous cascade (Chap. 14) with a 7×35. They have the added advantage that they are useful for numerous terrestrial activities, ranging from bird-watching through horse racing to sailing.

Small: Above 42 mm but less than 60 mm, with a magnification of 7–12×. Although like any binocular, they benefit from being mounted, most people can handhold these satisfactorily, especially if the elbows are supported. They are ideal for using in a reclining chair for observing high-altitude objects. They show hundreds of objects that are not visible to the unaided eye. They will have a field of view of 5–8°, with the majority having less than 6.5°. Although not quite as portable, especially in the 50 mm size, as the smaller class, they have a similar range of non-astronomical uses. Because they have a larger aperture than the ultra-portables, they can take more magnification. Increased magnification means that the sky background will be darker but star images brighter, leading to better contrast. The improved contrast and increased magnification makes more objects visible, and more detail can be seen in those that are visible in smaller binoculars. Typical examples are 10×42 and 10×50; the latter is, with reason, often classed as the ideal

---

[4]For example, Zarenski, 2004

first astronomical binocular. However, a well-made 10×42 will show as much as a budget 10×50 and will be more portable and easier to hold steadily.

Medium: 60–80-mm aperture, with a magnification of 20× or less. Although binoculars of this size can be handheld briefly, when they will show more than can be seen in a small binocular, to be used effectively, they must be mounted. Because of their size, their use outside astronomy is limited, although some people do use them for aircraft spotting and, mounted, for coastal observation. Almost all binoculars in this category are "straight through," so you will need to be seated if you are to observe high-altitude objects in any sort of comfort. A typical, and almost ubiquitous, example is the 15×70. This size of binocular is a serious observing instrument in its own right, and it is worth investing in the wherewithal to enable you to exploit its capabilities, which include a thousand or more objects that you cannot see with the unaided eye. The field of view is more restricted, typically around 4–4.5°.

Binoculars of this class come in a variety of qualities, from the very inexpensive "budget" ones through to what some people consider to be one of the best astronomical binoculars ever made, the 22×60 Takahashi *Astronomer*. The phenomenon of "you get what you pay for" is possibly more pronounced in this size of binocular than in any other. This is due in part to the development of extremely cheap 15×70s leading to competition that has driven prices down at the budget end. The trade-off is that the budget ones seem to have a very poor specification and almost nonexistent quality control (see section "Budget versus Quality").

I have found that the ideal mount is a monopod with a trigger-grip ball-head. This maintains the portability that is an attraction of binoculars of this size but renders it far more effective and makes it a pleasure to use. My monopod-mounted Helios Apollo 15×70 has become my favorite "grab-and-go" instrument (Fig. 3.1).

Large: Aperture greater than 80 mm but less than 110 mm, with a magnification of 20× or more. Binoculars of this size have to be mounted, so they are really out of the realm of "grab-and-go" and firmly in that of "serious observing kit." I consider them to be too heavy for a monopod, so a tripod, pier, or dedicated observing chair is essential. It is worth considering acquiring those with angled eyepieces as these make the observing experience, especially of high-altitude objects, far more pleasurable. Some have interchangeable eyepieces, enabling you to vary the magnification. They have limited fields of view, usually considerably less than 3°, so some sort of finder is very desirable—a unit-power reflex sight is ideal, being lightweight and better than adequate. They potentially show you tens of thousands of objects that are unavailable to the naked eye. Typical examples are the 25×100 and the 20/37×100.

Their relative portability, ease and speed of setup, and tremendous versatility have made my large binocular my most-used observing instrument. I find it ideal for public observing events, because most people find a binocular instinctive to use, so eye placement is less of a problem than with a telescope, and the aperture makes many of the "showpiece" deep-sky objects very easy to see.

Fig. 3.1 Monopod-mounted 15×70

Anything larger than this really has to be considered to be a "binocular telescope."

Figure 3.2 gives a comparison of what may be seen of the open cluster M35 through different classes of binocular under a darkish suburban sky.

Field of View

In addition to the magnification and aperture, the other numerical factor that is usually stated is the field of view. This is quoted in one of three ways:

- **Degrees**. This is the most useful one for astronomers, since it gives you an indication of the amount of sky that you will be able to see. The area of sky that will be visible is directly proportional to the square of the angular field.

Field of View

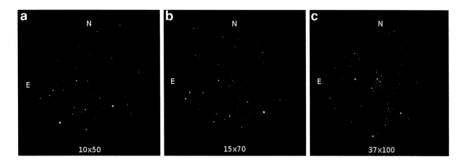

Fig. 3.2 Size matters: this is the open cluster M35 (Gemini) with (**a**) 10×50, (**b**) 15×70, and (**c**) 37×100 under good suburban skies

- **Meters at 1,000 m**. This is more useful for terrestrial use and is nowadays the most commonly found alternative to degrees. The approximate conversion of this to degrees is to divide by 17.5. Thus, 87 m at 1,000 m = $(87/17.45)° = 5°$.
- **Feet at 1,000 Yards**. This is more useful for terrestrial use and is nowadays most often found on binoculars intended for the US market. The approximate conversion of this to degrees is to divide by 52. Thus, 364 ft at 1,000 yd = $(364/52)° = 7°$.

Most, but not all, people prefer a wide field of view for astronomy. The true field of view is dependent on the magnification and the apparent field of view of the eyepiece:

True Field = Apparent field ÷ Magnification

Strictly speaking, an eyepiece can have an extremely large field, but this deteriorates rapidly towards the edge, so is limited by a field stop. There is always a trade-off between field of view and edge quality. In general, a 50° apparent field is a "standard" field, 65° and above is considered to be "wide angle," and 80° and above is designated "ultrawide angle." By comparison, the field of view of the unaided eyes is approximately 65°. Some manufacturers tend to be "optimistic" in their stated fields of view. In practice, 65° appears to be the upper limit for an apparent field; all binoculars I have used with wider apparent fields have suffered from severe deterioration of quality and easily noticeable vignetting in the outer part of the field, and those of ultrawide angle have also appeared to have a poorer image quality even in the center of the field when compared to standard field binoculars of a similar price.

Another problem associated with some very wide-field eyepieces is the effect that is colloquially called "kidney beaning" or "flying shadows." The colloquial names are descriptive of what you see if your eyepieces are afflicted with this problem, the correct name for which is *spherical aberration of the exit pupil*.

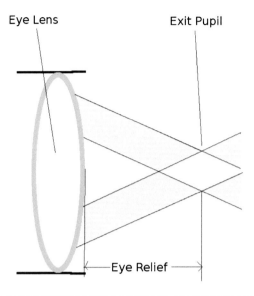

Fig. 3.3 Eye relief

Different zones of the exit pupil are focused at different distances from the eyepiece, so your eye is unable to focus on the entire field at once. If your eye is slightly off-center, the result is these *flying shadows* that are the shape of *kidney beans*. They tend to be worse at night when your pupils are more dilated, and some people seem to be more bothered by them than do others.

While wide-field views are an attraction to many people (a 7° field shows an area of sky twice as large as a 5° field), magnification is an extremely important factor for binocular astronomy, and a small cluster, nebula, or galaxy that is detectable at ×10 may appear to be stellar at ×7.

Eye Relief

The eye relief of a binocular is the distance from the eyepiece that you need to place your eye in order for all the light from the eyepiece to pass into your eye when the exit pupil of the binocular is the same size as the pupil of your eye. It is measured from the back surface of the eye lens (Fig. 3.3). It is the position of what physics textbooks call the "eye ring" and can be defined as the position of the image that the eyepiece forms of the objective lens. At this distance, you will be able to see the entire field of view and will have the brightest possible image. Over the last decade or so, manufacturers have become more aware that spectacle wearers will seek out binoculars with adequate eye relief to enable them to see the whole field of view.

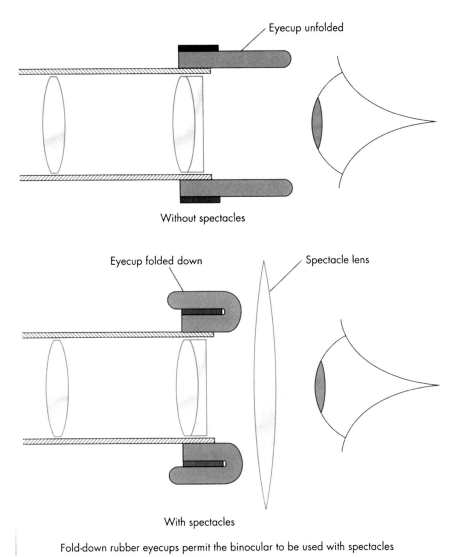

Fig. 3.4 Eyecups

As with fields of view, several manufacturers tend to be somewhat "optimistic" in their quoted eye relief, so, if you need to wear spectacles for observing, you should verify *in practice* that the binoculars have a suitable eye relief. In order that binoculars can be used by people both with and without spectacles, they will have eyecups that are either twist-down or fold-down to enable the correct positioning of your eye (Fig. 3.4).

Fig. 3.5 Fold-down eyecups are either fully up or down

Fig. 3.6 Twist-up eyecups can be used at an intermediate extension

Fold-down eyecups must be either fully up or fully down. Twist-up eyecups may be fitted with click-stops so that they can be used at any extension between fully up and full down, making them more versatile (Figs. 3.5 and 3.6).

As eye relief increases, it becomes increasingly difficult to position your eye precisely behind the eyepieces. This can exacerbate the kidney-bean effect if it is present.

Handheld Binoculars

Handheld binoculars are the choice for extreme portability, casual observing, and as a preliminary "sky-scanner" used in conjunction with a larger instrument. Almost all binoculars, including cheap plastic opera glasses, will show you more than the naked eye, but a sensible lower limit of aperture for portable astronomical binoculars is 30 mm. If extreme portability is not an issue, 40 mm is significantly better as it will admit more than 75 % more light. As aperture increases, so does the weight of the binocular, making it increasingly difficult to hold steadily. The sensible upper limit of aperture for hand-holding is normally considered to be 50 mm. Larger apertures than this can be handheld for short periods but are too tiring to use for anything other than very brief views.

For many years the "common sense" view was that the limit to magnification for handheld 50-mm binoculars was ×7. This is probably because 7×50 was, and still is, the most common size of handheld marine binoculars. While it is true that they are easier to hold steadily than, say, 10×50, most of us do not do our astronomy from the moving deck of a boat. There are very few astronomical objects that are better at 7×50 than at 10×50, and it is perfectly possible to hold 10×50 binoculars sufficiently steadily when our observing platform is the ground (see Chap. 6). The increased magnification of the 10×50 allows us to see more detail and generally gives more satisfying views. This is reflected in its higher rating in every performance index except relative brightness, and even this difference is reduced for those of us whose pupils do not dilate sufficiently to enable us to use the full 7-mm exit pupil of the 7×50.

There is perennial debate on whether, if you use small- or medium-sized binoculars for astronomy, you should use Porro-prism or roof-prism binoculars. The conventional wisdom is that Porro-prism binoculars are better for astronomy. While it is certainly true that *for the same price* Porro-prism ones tend to have superior optical quality, good-quality roof-prism binoculars are optically and mechanically good as good-quality Porro-prism binoculars. The roof prism ones tend to be lighter and more compact and are therefore generally easier and more comfortable to hold. Over the recent years, I have found that I use my 10×42 roof prisms more than my 10×50 Porro prisms and a side-by-side comparison shows that I see no more in the Porros when I handhold them, although they do show very slightly more when they are mounted. Roof-prism binoculars also have the advantage that they can more easily be made waterproof, as the focusing mechanism is usually internal.

If you choose Porro-prism binoculars, you may also have a choice between center-focus and independent eyepiece focus. There are no advantages to center-focus if the binoculars are to be used exclusively for astronomy, but, if you intend to use them for terrestrial purposes (e.g., bird-watching or horse racing), then you

Fig. 3.7 Variety of 10×50 Porro-prism binoculars. L to R: older-style center-focus Z-body with fixed eyecups (*Zenith*); robust center-focus B-body style (*Swift Newport*); modern lightweight center-focus Z-body with folding eyecups (*Helios Naturesport*); robust individual focus Z-body (*Strathspey Marine*)

Fig. 3.8 Canon 10×30 IS image-stabilized binoculars (Photo: Canon Inc.)

should get center-focus. My preference for astronomy is independent focusing eyepieces. They tend to be more mechanically robust and do not suffer from a rocking bridge, and modern ones tend to be waterproof and nitrogen filled, reducing the likelihood of internal fogging (Fig. 3.7).

Another option for handheld binoculars is those with image stabilization (Fig. 3.8). These incorporate an electronic system that compensates for motion and vibration. Different manufacturers employ different stabilization systems, which were developed initially for military surveillance and not for astronomy. A test report in *Sky & Telescope*[5] suggests that the best stabilization system for astronomical observation is that employed by *Canon*, whose optics were also superior. In addition to the stabilization system, the optics are essentially a roof-prism system with a field flattener (see Chap. 2) incorporated into the design. The stabilization system (see Chap. 2) compensates for shake and the result is that you can see fainter objects and more detail. A 10×30 IS binocular will show most people more than a conventional 10×50. While a 10×30 IS is sufficiently light (600 g/1¼ lbs) to be handheld for relatively long periods, the larger 15×50 IS and 18×50 IS are heavy enough to be tiring to hold for extended periods. If you are considering purchasing image-stabilized binoculars, you should be aware that the image stabilization mechanism requires battery power and that the life of batteries, particularly alkaline batteries, is reduced in cold weather. The development of reliable high energy density, battery technology in the last decade has made these a much more attractive option. Without power, the binoculars can be used as conventional binoculars.

Mounted Binoculars

All binoculars will show more if they are mounted because this eliminates the "jiggles" of hand-holding. Even a small binocular will show objects a magnitude or so fainter if it is mounted. If binoculars are going to be a main observing instrument, it makes sense to acquire ones of greater aperture and magnification than can be handheld. Big binoculars in the aperture range 60–100 mm have become readily available in recent years (Fig. 3.9).

Once you are dealing with big binoculars, you are dealing with specialist instruments and can expect them to have features that enhance the ease and pleasure of astronomical observing. These may include:

Mounting Plate. Because big binoculars necessarily have to be mounted, it is common for them to incorporate a plate, with 1/4-inch UNC threaded holes, for direct mounting to a photographic tripod or other mounting. This eliminates the need for an L-bracket, which inevitably introduces an additional potential source of instability and is yet another essential piece of equipment that has to be remembered (Fig. 3.10).

Angled Eyepieces. There is little to recommend straight-through binoculars for astronomical observing. They are considerably less comfortable than those with angled eyepieces when you are observing at high elevations. Angled eyepieces also permit the use of photographic or video tripods and heads to be used because they eliminate the need to "limbo dance" under the tripod when you observe objects of high elevation.

---

[5]Seronik 2000

Fig. 3.9 Strathspey 15×70 binocular. These low-priced Chinese binoculars, which are sold with different brand names, have become very popular in recent years (Photo: John G. Burns)

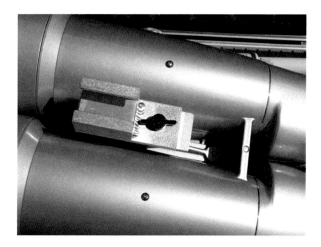

Fig. 3.10 Bracket attached to mounting plate at the bottom of 20/37×100 binoculars

Interchangeable Eyepieces. If binoculars are mounted, interchangeable eyepieces become functionally useful. The ability to change magnification permits, within the limits of the mechanical and optical precision of the binocular, the best combination of image brightness and contrast to be selected. Interchangeable eyepieces are usually a friction-fit into the eyepiece holder, although some binoculars have their eyepieces turret-mounted so that the unused eyepieces cannot get mislaid or dropped. I am not convinced that this is a long-term advantage, because it introduces another feature that has to be made to great precision and detracts from the inherent simplicity of binoculars by adding to the number of things that can go wrong. On the other hand, their advocates report this feature to be extremely useful.

Budget Versus Quality

The last few decades has seen an influx of binoculars from China, many of which are retailed at an exceptionally low price. This represents a remarkable achievement, which has made this class of binocular available to many more people than was the case 15 years ago when you would have needed to pay ten times as much for a 70-mm astronomical binocular (albeit of far better quality). However, this comes at a price, which is made apparent by this class of binoculars being over-represented in "how do I solve this problem?" type threads on Internet astronomical forums. Think about it: it is not uncommon, for example, to find a 15×70 binocular retailing for less than a pair of reasonable quality astronomical eyepieces. A binocular has two eyepieces, two objectives, two different focusing mechanisms, prisms and housing, and other bits of associated hardware. Realistically, what sort of quality binocular is it reasonable to expect for the cost of one and a half eyepieces (in the case of the 15×70)?

Many people claim to be satisfied with these binoculars and undoubtedly some are. It is also the case that we find it psychologically difficult to admit to a poor choice, so we tend to fool ourselves; it is a common, and perfectly normal, human trait.

There are several major issues with these binoculars. Firstly, and possibly most important, is that quality control is poor to the point of being almost nonexistent. Secondly, the prisms have a tendency to be knocked out of alignment very easily. Thirdly, in order to reduce the severity of aberrations, the aperture is effectively stopped down by internal baffling. Measuring the exit pupil of some of these binoculars has revealed that some 50-mm binoculars actually have an effective aperture as low as 42 mm and that the ubiquitous 15×70 may have an effective aperture as low as 62 mm. This is what one of the manufacturers says[6]:

---

[6]http://www.binocularschina.com/binoculars/MS.html

> *For years, the international markets are flooded with unbelievably low-price Chinese binoculars and some of the users have been complaining or bashing loudly about the quality control consistency of Chinese binoculars for a while. Actually, it's quite simple to improve the quality consistency: spending much more time in grinding and selecting glass, spending much more time in training the workers for assembling the binoculars, and spending much more time in the final quality check - then, a much better quality binocular will be made, however, at the trade-off of much higher production cost.*

This confirms a long-held suspicion that the manufacturer finds it more cost-effective to let the customer do the quality control on these budget offerings. An unknowledgeable customer is more likely to accept a binocular of such poor quality that any self-respecting quality controller would have to reject it, and the company only effectively pays for the quality control of rejected binoculars.

A common feature of these budget offerings is that, as a consequence of the way the prisms are mounted, they lose collimation very easily. A well-made binocular will hold collimation for decades unless it is severely abused (usually to the extent that the outer casing is dented or shows other signs of abuse); a high proportion of the budget binoculars reach the customer already out of collimation. Poor collimation may result in double images or, if it is of lesser severity, eyestrain that can lead to headaches and/or nausea. Many users are prepared to accept this and learn to realign the prisms themselves. This is really an essential skill to acquire if you have one of these binoculars. It is daunting at first, but becomes intuitive with practice (and some binoculars provide plenty of this!).

A less common, but still notable, feature of these is that the right eyepiece diopter may be poorly set, resulting in a user being able to focus for both eyes. In some models, this can be very easy to remedy (see Chap. 5).

At the other extreme of Chinese binoculars is some very good-quality instruments. While these are not quite the quality of binoculars like, for example, the Fujinons, Swarovskis, or Leicas, they are not far off it and cost a fraction of the price. In my opinion, these better-quality Chinese binoculars are exceptional value for money. They tend to be robustly made and, optically and mechanically, are very good indeed. The 15×70, for example, is nearly a magnitude brighter than the budget equivalent, has much better coatings, and is nitrogen filled. Mine has become my most-used grab-and-go binocular.

Binoviewers

Binoviewers (Fig. 3.11) are designed to permit the use of two eyes with a single optical tube assembly (Figs. 3.12 and 3.13). The rationale for their use is that they offer some of the advantages of binoculars with few attendant disadvantages. The obvious advantages are the reduction in eyestrain from using two eyes, the suppression of the blind spot, and the aesthetics of false stereopsis (see Chap. 1). The obvious disadvantage is the loss of light into each eye that results from the splitting of the light into two optical paths and from the additional optical elements in each light path. While binocular summation (see Chap. 1) can compensate for some of

Fig. 3.11 Skywatcher binoviewer with two pairs of matching eyepieces and a nosepiece-fitting Barlow lens

Fig. 3.12 Denkmeier binoviewer used with a limited edition 12.5 in. f/6 Teeter's Telescopes "Planet-Killer" telescope (Photo Copyright 2005, Teeter's Telescopes)

Fig. 3.13 Celestron binoviewer attached to a Meade 10″ LX50 (Photo courtesy of Gordon Nason)

this loss, the overall perception is that using a binoviewer is equivalent to a loss of about one third of the illumination as compared to a single eyepiece. In addition, the cost of providing matching eyepiece pairs, thus doubling the number of eyepieces required when compared to a conventional telescope, is not one that can be ignored, especially where good-quality eyepieces are used. However, this is ameliorated to some extent by the development of binoviewers that incorporate the facility of multiple magnifications without changing the eyepieces.

An advantage that is not common to conventional binoculars is the ability to use high magnifications without the need to collimate two optical tubes. Another is that their use with telescopes of large aperture provides an equivalent aperture that would be significantly more expensive and technologically difficult to achieve with binoculars or binocular telescopes. A less obvious advantage when they are used at high powers concerns "floaters." Floaters are strands of protein that float within the transparent humors of the eye and which become apparent, sometimes distractingly so, when one is observing with a small exit pupil. Users of binoviewers report that their visibility is suppressed, often to the point of elimination, probably in the same way as they suppress the blind spot.

If you are considering a binoviewer, you should ascertain its clear aperture, as this will place a limit on the lowest power of eyepiece that you can use effectively. In cheaper units, this can be as little as 20 mm, restricting the eyepieces to those with a field stop less than or equal to this. You should also ascertain whether your telescope has sufficient back focus to permit it to be used with a binoviewer; a Barlow lens in the "nose" of the binoviewer may enable this. Finally, you should consider those that have self-centering eyepiece holders as this will eliminate any miscollimation that may otherwise result from slightly undersized eyepiece barrels being held off-center by thumbscrews.

Zoom Binoculars

The question of zoom binoculars is one that inevitably arises, not least because there are good-quality zoom *telescopes* and good-quality astronomical zoom *eyepieces* on the market. I once made the comment that a decent zoom binocular for astronomy had yet to be invented. The vastly experienced binocular repair man, Bill Cook, retorted to the effect that my qualification "for astronomy" was redundant. The reasons for this are simple. Not only must the eyepieces zoom to within 1 % of exactly the same rate (which means absolutely no perceptible rocking of the bridge), but a zoom binocular requires a system with moveable optical elements that must hold collimation, ideally to better than an arc minute where step (aka dipvergence, aka supravergence) is concerned if one is approaching ×30; for the ×125 that I have seen advertised for some zoom binoculars, this translates to better than 15 arcseconds! Now, consider how many good-quality center-focus 30× binoculars you know of—I don't know of any, and I am sure that part of the reason must be that it would be a feat of technological brilliance (not to say expense!) to bridge two eyepieces in such a way that they maintain collimation to within the tolerances that are required. (And remember that it is unlikely that they will have a "base tolerance" of zero error.)

According to Seyfried, zoom binoculars were developed as a "gimmick to stimulate sales" on the back of the success of zoom lenses for cameras. He also asserts that the frequency with which they fail results in their being disproportionately represented at binocular repair facilities.[7] He also states that he has never seen a zoom binocular that can hold collimation.[8]

Bibliography

Adler, A., *Some Thoughts on Choosing and Using Binoculars for Astronomy*, Sky and Telescope, Vol.104 No.3, September 2002, pp 94–98

Bishop, R., **Binoculars**; in Gupta, R.(ed.), *Observer's Handbook 2003*, Toronto, University of Toronto Press, 2002, ISBN 0-9689141-2-8, ISSN 0080–4193

Seronik, G., *Image-Stabilized Binoculars Aplenty*, Sky and Telescope, Vol.100 No.1, July 2000, pp 59–64

Seyfried, J.W., **Choosing, Using, & Repairing Binoculars**, Ann Arbor, University Optics Inc., 1995, ISBN 09346*39019*.

Zarenski, E., *How-to Understand Binocular Performance,* http://www.cloudynights.com/documents/performance.pdf

---

[7]Seyfried, p. 10ff
[8]Ibid, p. 47

Chapter 4

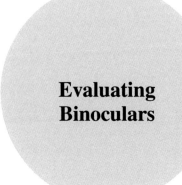

Evaluating Binoculars

A few years ago, a colleague told us of a local petrol station that was offering 10×22 compact binoculars for sale at £2.99 each. On the reasoning that "you can't go wrong at that price," the colleague acquired a pair for his son, who was very pleased with them. Upon hearing of this, another colleague went to the petrol station and bought a pair for herself. When she got home and tried them out, she found that they gave a double image, obviously a case of poor collimation. She returned them to the petrol station, where the sales assistant took them and, without even checking them, placed them in a box and replaced them with a new pair out of another box. My colleague was satisfied with the replacement pair and pleased with the "service" she received from the petrol station.

A few weeks later, I took a group of pupils to see an international cricket match; one of them was the son of the first colleague, who brought his new binoculars with him. As usually happens, these binoculars were passed around among the pupils, most of whom had never used binoculars at a sporting event, were impressed with the magnified image, and wanted repeated looks through the binoculars. To ease the demand, I passed around my good (but by no means superb) quality 10×42. Every pupil immediately noticed the difference and it was obvious that none had used binoculars of this quality before. As one put it: "These are amazing, Mr. Tonkin. They are even clearer than eyesight!" The colleague's son was, as you can imagine, a bit deflated because his binoculars seemed so inferior. I pointed out that mine had cost almost a hundred times the cost of his. I showed the pupils how to detect the off-axis chromatic aberration and pincushion distortion in mine and then asked them to consider if they thought that the image in mine was a hundred times better. I also pointed out that mine could not be conveniently carried in a shirt pocket. Honor was satisfied and we got on with enjoying the match, albeit with my binoculars having far more use than the budget pair.

This pair of episodes illustrates at least half a dozen things:

- Over the last few decades, binocular manufacturing methods have improved to the extent that, without any but the most rudimentary quality control checks, binoculars of reasonable quality can be produced remarkably cheaply. Budget quality binoculars can be produced so cheaply that they are effectively disposable items.
- It is far more cost-effective for manufacturers of budget-quality binoculars to use the customer to do the quality control. It is cheaper merely to replace the unsatisfactory (to the customer) instruments than it is to employ quality control staff.
- People will tend to be satisfied with poor quality unless they have something better with which to compare it. The consequence of this is that many instruments, which may have been rejected by effective quality control, will be acceptable to some customers, particularly if "the price is right."
- Differences in optical quality can often be immediately apparent, even to "untrained" people, most of whom are capable of performing simple tests for common aberrations.
- Once you have used good-quality binoculars, it is difficult to be satisfied with less. However, we do become emotionally attached to our possessions and can readily justify poor quality on the grounds of price or of some other comparative benefit like ultra-portability.
- Optics that are entirely free of aberrations exist only in the imagination and, to a large extent, the old adage that "you get what you pay for" still holds true. For recreational use, the determination of whether the extra quality is worth the extra price is almost entirely a subjective one.

Hence, it is not only possible but also very desirable to be able to do some initial testing of binoculars in the store where they are bought. With the advent of the phenomenon that an ever-increasing number of goods are bought, for reasons of cost, over the Internet or by mail order, the same applies to testing on receipt of the item. However, as you are aware, there is no substitute for the more demanding tests that astronomical use makes of binoculars, so it is important that you ensure that the vendor has a return policy that will permit you to return them if they are unsatisfactory when they are used under the stars.

Preliminary Tests

A very large amount of information can be gleaned from some very basic preliminary testing. This will eliminate binoculars that are grossly unsuitable. Remember that, with very few exceptions, aberrations, faults, or features that are merely irritating during initial testing will become infuriating under the stars. However, while you should not expect to find a binocular that is entirely free of all aberrations or faults, you should expect to find that they exist in lesser number and severity in more expensive instruments. What you can expect to have to accept depends to a large extent on your budget, and you may find that you are more sensitive to some aberrations or faults than to others. Your final choice must inevitably be a subjective one, but will ideally be one that is guided by a measure of objectivity.

With all tests that use touch or hearing, remember that closing your eyes tends to make these senses more acute for many people.

Firstly, do not even consider fixed focus, zoom, or "quick focus" binoculars; they are unsuitable for astronomy.

- **Visual Overview.** Reject any binoculars that have "ruby" optical coatings, loose screws or screws with damaged heads, covering material that is not in complete close contact with the binocular housing, scuff marks anywhere, any evidence of dust or other foreign matter on the inside of the optics, or internal baffling that is not uniformly matt black.
- **Mechanical Overview.** Give the binoculars a good shake. Reject any in which you feel or hear any movement of components.
- **Focus Mechanism.** Run the focus mechanism through its full range. The feel should be uniform throughout the range and should not be too stiff or too loose. If it is stiff, it is very difficult to *find* a precise focus. If it is loose, it is difficult to *maintain* a good focus. Feel for any stiff regions or any "sloppy" regions. Different feel in different regions is indicative of poor mechanical tolerances in manufacture. Feel for any difference between dynamic friction and static friction ("stiction") by stopping the focus at various points throughout the focal range and feeling for a slight jerk or "catch" when you start to refocus. This is usually due to poor-quality lubricant and makes precise focusing difficult. On binoculars that have individual eyepiece focusing, test each eyepiece focus individually. On center-focus eyepieces, remember to test the diopter-adjustment ring of the right-hand eyepiece.
- **The Bridge.** The bridge is the pair of arms that connects the eyepieces to the center-post focus mechanism in center-focus Porro-prism binoculars. All bridges will rock to some extent and, as they do so, they change the focus of either or both sides of the binocular. On well-made binoculars, the rocking is minimal and requires considerable pressure, more than will be put on the eyepiece housing in normal use. On budget-quality binoculars, there can be considerable rocking with minimal pressure. This rocking gets worse with age. To test the bridge for rocking, merely hold the binoculars by the prism housings with the eyepieces downwards, and press down alternately on the eyepiece housings with the tips of the forefingers. The severity of any rocking becomes immediately apparent. If you are unsure how it will affect you, hold the binoculars to your eyes, focus on something, and, by rocking the binoculars from side to side, put slight pressure alternately on each eyepiece housing with your eye socket. If the focus changes, reject the binoculars.
- **Interpupillary Distance (IPD) Adjustment.** In handheld binoculars, this is usually the central hinge. For larger binoculars, this can be either a hinge or eccentrically rotating prism housings or eyepiece turrets. If it is loose, it is difficult to maintain any given IPD. If it is very stiff, or if it is jerky, it is difficult

Fig. 4.1 Wide-angle eyepieces (*top*) are of large diameter and may be unusable for people with narrow-set eyes or wide noses

to set the IPD. If only one person is to use the binoculars at any one time, this need not be a significant problem. Be aware that, with several center-hinge binoculars, they need to be folded to near the minimum IPD in order to fit into the case; test if this is necessary with your IPD. Check that the IPD range accommodates all intended users. Most binoculars do not cover the entire range of IPDs for adults; this range is usually considered to be about 43–80 mm with a mean at around 65 mm and about 90 % of people having IPDs within 8 mm of the mean. Most binoculars do not deviate more than about 10 mm from the mean. If you know your IPD, the binocular IPD adjustment is trivially easy to check merely by using a piece of card with your IPD marked on it and holding it over the binocular eyepieces. If not, you can check this when you test the binoculars for optical quality (below). Check that the eyepieces go comfortably to your eyes when the IPD is set for you. If you have narrow-set or deep-set eyes, or if your nose has a wide bridge, this may not be possible for you especially with wide-angle eyepieces, which are typically larger in diameter than "normal" ones (Fig. 4.1).

- **Tripod Bush.** You will be able to see significantly more with mounted binoculars than with unmounted ones, even if you do intend to use them primarily as a handheld instrument. Most medium-sized modern binoculars have a 1/4-inch UNC bush in the distal end of the center post. This bush is covered by a cap, usually of plastic, but sometimes of metal, which unscrews. Remove it and

check the quality of the bush. If possible, try an L-bracket and ensure that you can easily connect the bracket to the binocular, with the thread of the L-bracket bolt easily meshing with the thread in the bush without danger of cross-threading. Remember that you will most likely be wanting to do this in the dark, possibly with cold or gloved hands.
- **Prisms.** Hold the binoculars away from you, pointing towards something relatively bright, e.g., the sky or a light-colored wall or ceiling (*not* the Sun!), and look at the bright circle of light in the eyepieces. If it is actually a circle, all well and good. If it is tending towards lozenge shaped, this is an indication of undersized prisms; undersized prisms are themselves indicative of cost cutting. If there are blue-grey segments of the circle with a brighter lozenge inside, this is indicative of cheaper BK7 glass in the prisms (Fig. 2.9). In both cases, the prisms will cause some vignetting of the image. With roof-prism binoculars, bring them up to the eyes and carefully examine the image of the bright surface. Do you see a faint line if you defocus your eyes? If so, you are seeing the "ridge" of the roof prism. If the line is obtrusive, it will result in flaring of bright astronomical objects. This is easier to test for using a bright point of light against a dark background, but this is usually not available in the store.
- **Eye Relief.** When you look through the binoculars, the image in the eyepiece should be surrounded by the crisp dark edge of a field stop at the mutual focus of the eyepiece and objective. If there is insufficient eye relief, you cannot get your eye sufficiently close to enable this. Most modern binoculars have fold-down or twist-down rubber eyecups around the eyepieces to enable their use by spectacle wearers (Fig. 3.4). However, not all rubber eyecups fold down, and not all permit a bespectacled observer to get his eyes sufficiently close. Even if you do not wear spectacles, do check that the eyecups fold or twist down. On cold or dewy night, warm moisture evaporating from your eye can condense on the eyepiece, causing it to fog. If you fold or twist the eyecup down, air can circulate between your eye and the eyepiece, reducing the likelihood of fogging.
- **Exit-Pupil Size.** Over the last decade, it has become apparent that some manufacturers of budget binoculars reduce the cost by effectively stopping down the aperture internally. This results in, say, a nominal 15×70 binocular actually being a 15×62 binocular, i.e., passing less than 80 % the amount of light than it would if it was the full aperture. Tests by amateurs in the astronomy and bird-watching communities have tested dozens of binoculars for both magnification and effective aperture. Whereas the magnification tends to fall well within the industry-standard tolerance of ±5 % of the stated value, apertures (for which, apparently, there is no industry-standard tolerance—presumably, when tolerances were agreed, nobody assumed that this sort of cost reduction would be employed) have been found to be as much as 20 % lower than the stated value! Assuming the magnification to be accurate, a rough idea of actual effective aperture can be achieved by measuring the exit pupil. A simple way of doing this is to use millimeter-ruled graph paper behind the eyepiece, when the binocular is focused at infinity and aimed at the daytime sky. The exit pupil is where the image of the illuminated objective is smallest and has the sharpest circumference.

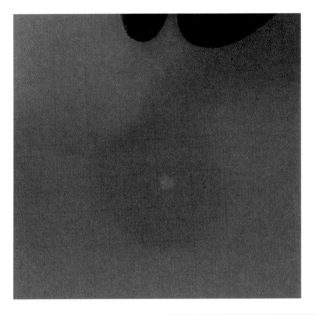

Fig. 4.2 The exit pupil of this nominally 10×50 binocular is clearly undersized, suggesting that the effective aperture is less than 50 mm

The exit pupil should be aperture/magnification, e.g., a 10×50 binocular should have an exit pupil of 50 mm/10 = 5 mm. If the magnification is within acceptable limits, it will be no smaller than 4.76 mm. The graph paper can measure the exit pupil to a precision of 0.5 mm, so any grossly undersized effective aperture will become immediately apparent (Fig. 4.2).

- **Comfort.** This is particularly important if you intend to use handheld binoculars for long periods of time. In general, lighter binoculars are less tiring to hold, but this is not always the case. A heavier binocular that is well designed from an ergonomic perspective can be less tiring than a lighter instrument that is poorly designed. There is no substitute for experiment when it comes to determining how a particular binocular suits you as an individual. When you perform the optical tests below, take your time and perform them consecutively without taking the binoculars from your eyes. Be conscious of how tired your arms become.
- **Focus.** (See also *Spherical Aberration*, below.) To focus a binocular, do one optical tube at a time, with the other side covered with a lens cap on the objective side. First of all, set the interpupillary distance. Check, by alternately shutting eyes or alternately covering objective lenses, that you have a complete field of view, surrounded by the field stop, on both sides. If the eyepieces focus independently, it does not matter in which order you do them. With center-focus binoculars, cap the right-hand objective and focus the left-hand optical tube on a distant object with the focus wheel (called "focus band" in some binocular instruction

sheets). Critically examine the image and try to determine if it "snaps" to a good focus or if there is a range where it looks less "mushy." When you have best focus, swap the lens cap to the left objective and, *without adjusting the focus wheel*, focus the right-hand tube with the diopter-adjustment ring on the right-hand eyepiece housing. Again, critically examine the focus. Remove the lens cap, and again examine the focus. Then focus on a nearer object and, by alternately covering the objective lenses, verify that both sides are focused, not merely the one used by your dominant eye. Do not be tempted to use a hand instead of a lens cap or, worse, to focus individual tubes by closing one eye; this is rarely satisfactory. The hand almost always changes the mutual orientation of eyes and binoculars from the usual observing position. This is true also when the binocular is mounted, merely because of the act of stretching the arm to a non-observing position. Closing one eye can cause the other to squint. When you do visual optical tests, you want your eyes and body to be relaxed.

- **Focal Range.** When binoculars are used for astronomy, it is not immediately apparent why one would need any focal distance other than infinity. There is, of course, the trivial case of the applicability of binoculars to terrestrial use, which often requires that you can focus them *closer* than infinity. The other case is that of spectacle (eyeglass) wearers who wish to observe without spectacles. If your eyes are hypermetropic (longsighted), there is usually not a problem because the binocular adjustment is to what would be focus on a nearer object for a person with normal vision. However, if your eyes are myopic (shortsighted), to focus on an object at infinity, the binocular must be adjusted to focus on what, to a person with normal vision, would be *beyond* infinity. Whereas most binoculars have some facility for this, the amount of extra focus travel varies enormously. If your eyes are myopic and you wish to observe without your spectacles, you should verify that the binoculars have sufficient focal range to accommodate your eyesight. You should try to focus on an object at as great a distance as possible—at least half a mile, but preferably more—away from you. Additionally, because the depth of focus of your eyes is reduced when your pupils are dilated, as they will be when the binoculars are used for astronomy, you should try to focus on a dark object without a bright background. Dark vegetation on the horizon is ideal. Be aware that you will almost certainly need a bit more "beyond infinity" travel at nighttime than you need during the day, so allow for this.
- **Internal Reflections.** Internal reflections, be they reflections off the interior walls or components of the binoculars, or "ghost" reflections off poorly designed or inadequately coated optical components, are both distracting and detrimental to astronomical observation. They tend to be most obtrusive when a small bright object is observed, off-axis, against a dark background, i.e., exactly the conditions that are often found in astronomical observation! To test for this, use a bright light source (*not* the Sun), such as a recessed halogen lamp in the store ceiling or the LED camera flash of a mobile phone, slightly off-axis.
- **Field Curvature.** Lenses do not focus images on a plane, but on a curved surface that is concave towards the lens. The result is that if the eyepiece is focused on an object at the center of the field of view, objects at the edge will be out of focus.

This is field curvature and it is potentially present in all binoculars. In excellent binoculars it may not be noticeable at all; in budget binoculars it can be obtrusive less than halfway to the edge of the field. Unless it is severe, it is not a major problem for daytime use, where the attention is on objects at the center of the field of view, but astronomers prefer pinpoint star images right to the edge. It can be ameliorated by addition of extra lens elements, and the field of view can be limited by field stops so that it is not apparent. The extra lens elements will absorb some light and reduce contrast, an important consideration for astronomical use. In ultrawide-angle binoculars, unless they are extremely well designed, it can render most of the field of view unusable for astronomy, thus negating the perceived advantage of a wide apparent field of view. To test for it, merely focus on an object at the center of the field of view and move the binoculars so that it moves towards the edge. If it goes out of focus but can be refocused, this is field curvature. (If it can't be refocused, it is probably coma.)

- **Chromatic Aberration.** All binoculars will exhibit some degree of chromatic aberration. In very good binoculars it may only be noticed off-axis and then only just perceptible. The purpose of this test is to compare binoculars, not to find one that is perfectly achromatic. Chromatic aberration is most obtrusive with high-contrast objects, such as many astronomical targets. Although it may not be noticeable on fainter or lower-contrast objects, if it is present, it *will* degrade the image by reducing contrast. The simplest daytime test is usually to view a distant TV antenna or electricity pylon or similar against a bright sky. It is important to ensure that the IPD is properly set, since chromatic aberration can often be induced by moving the eye off-axis. Focus the target object at the center of the field and slowly pan the binoculars so that the object moves towards the edge. Chromatic aberration will be visible as colored fringes at the interface of light and dark, usually magenta on one side and cyan on the other.
- **Spherical Aberration.** (See also *Focus*, above.) Spherical aberration occurs when light from different regions of the lens is focused at different distances from the lens. It is usually well corrected for in modern binoculars but does exist in budget ones. It is visible as a "mushy" focus, i.e., an object at the center of the field of view does not "snap" to a good focus.
- **Coma.** Coma is a form of off-axis spherical aberration. It is very difficult to test for during the day as it requires a point source of light. This is sometimes possible by viewing a glint of sunlight reflected off a very curved shiny object such as a metal car radio antenna. Focus the glint in the center of the field and move the binoculars so that the object moves towards the edge. If coma is present, the image of the glint will flare towards the edge of the field, giving it the appearance of a comet (hence the name "coma") with its head towards the center. Coma is often present in binoculars that are not specifically designed for astronomy. This is because it is not normally visible or particularly degrading of daylight images, largely because birders, for example, use binoculars to examine birds at the *center* of the field.
- **Astigmatism.** Astigmatism is an aberration that, like coma, is very difficult to test for during the day. It manifests itself as a point object, such as a glint of

sunlight, being seen as a short line when just out of best focus, that changes orientation through 90º from one side of focus to the other.
- **Vignetting.** Vignetting results from the outer parts of the field of view not being illuminated by the whole of the objective lens. It manifests itself as a darkening towards the edge of the image. It is present in all terrestrial binoculars, where it is not obtrusive because of their use to examine objects at the center of the field of view, and most astronomical binoculars. Mild vignetting can be difficult to test for during daylight because the human visual system readily adapts to a very large range of illumination, but flicking the gaze back and forth between edge and center of the field of view will usually reveal it.
- **Kidney-Bean Effect.** The kidney-bean effect, also known as *flying shadows*, is an affliction associated with some wide-field eyepieces and is a result of spherical aberration of the exit pupil. Instead of being a flat disc, the exit pupil is curved and it is therefore impossible to focus the entire field of view at once and you have to hold the binocular slightly further from the eye to focus one zone than another. There is a position at which your iris will cut off the light from a zone between the center and the periphery, and, if your eye is not perfectly aligned with the optical axis of the eyepiece, the result is these flying shadows that have the shape of a kidney bean. The effect is more pronounced in daylight, or when viewing a bright Moon, than at night and is therefore best tested for during daylight (Fig. 4.3).
- **Distortion.** Almost all binoculars will exhibit some distortion towards the edge of the field. To test for it, focus on a straight object such as a telegraph pole or a roof ridge that extends across the diameter of the field of view. Move the binoculars so that the edge of the object forms a chord near the periphery of the field. If the object curves inwards towards the middle, you have pincushion distortion; if it curves outwards, you have barrel distortion. A small amount of pincushion distortion can be desirable for terrestrial use, but it has no advantages—or significant disadvantages—for astronomy.

It is a common phenomenon, even among experienced binocular users, that, when they optically evaluate a binocular, they notice pincushion distortion and comment adversely on it. It is far less common (I hope that this book will go some way towards remedying this) that they know the whole reason. It is indeed true that pincushion distortion, which manifests itself as straight lines at the edge of the field of view appearing to curve in towards the middle of the field, results from increasing angular magnification away from the center of the field, but this difference in magnification is not an error, it is intended. If there is equal angular magnification, the linear magnification at the edge of the field is less than in the center, and an optical phenomenon called "rolling ball effect" occurs when the binocular is panned. This may not be noticeable when the binocular is used astronomically, but, when it occurs in terrestrial use, it can be distinctly unpleasant and can cause nausea. A small measure of pincushion distortion eliminates this rolling effect. Different manufacturers choose different compromises between "rolling ball" and pincushion; the choice of which particular compromise is best is entirely subjective.

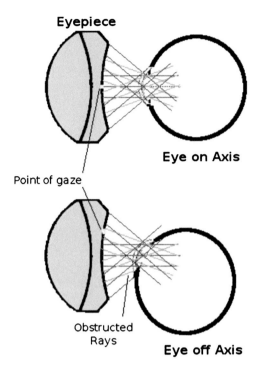

Fig. 4.3 When the eye is not perfectly on axis, with the eyepiece, the iris will block some peripheral rays and cause "kidney beaning"

- **Collimation.** Unless miscollimation is severe, this is usually considered to be the most difficult condition to test for. Unfortunately it present in a very large number of lower-priced binoculars. Severe cases manifest as a double image that cannot be got rid of. Our eyes can compensate for mild miscollimation, but the price is eyestrain that can lead to fatigue and headaches if it is prolonged. The acceptable limits for miscollimation are given on page 40, but these limits are not the ones that can be measured during a preliminary test. A daylight test for convergence and divergence is to support the binoculars and focus on a *distant* vertical target such as the edge of the wall of a distant building or a telegraph pole, and close one eye and place the target at the edge of the field of view. Alternately close your eyes and see if the image in one eye is laterally displaced from that in the other. To test for *step*, i.e., vertical shift, use a horizontal target such as the ridge of a distant roof. We are far less tolerant of step than we are of convergence or divergence, and, if you detect any step at all, you should reject the binoculars.

  Tests for miscollimation can be easier to perform if you can "fool" the brain into treating the images in each eye as if they are of different objects; it then does not try to make your eyes merge them if they are displaced. One simple way to

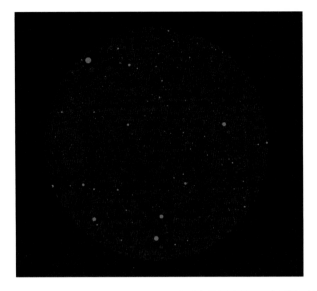

Fig. 4.4 Anaglyph glasses can cause the visual cortex to process the image in each eye as different objects

do this is to use a different colored filter on each side. The red/cyan glasses used for anaglyph 3D are ideal, but a similar effect can be achieved with different colored cellophane sweet/candy wrappers (making sure, of course, that you don't transfer any confectionery to the eyepieces!) This "fooling the eye" is easier to achieve at night, with bright stars than it is during the day with familiar objects (Fig. 4.4).

Field Tests

If binoculars are to be used for astronomy, there is no substitute for field-testing them under the stars. The night sky offers the potential of objective and quantitative testing that is not easily available in daylight without relatively sophisticated optical equipment. In addition, with the exception of distortion, all the optical tests detailed above are easier to perform under the night sky. You will obtain better, and more easily comparable, results if you mount the binoculars.

- **Overall Optical Performance.** You can perform a simple and, to some extent, quantifiable test of the optical quality of your binoculars by determining the closest double stars that you can distinguish. In general, double stars in which the components are of approximately the same magnitude are easier to split than those where the components are of significantly different magnitudes. The ability to separate double stars is obviously a function of magnification, so you should therefore expect binoculars with higher magnifications to routinely outperform those with lower magnifications. Because you are observing at low magnification, you should not expect to see the close separations that are discernible with telescopes of similar aperture working at high magnification. You should also be aware that differences between sky conditions at different times and places and differences in the optical acuity and observing experience of different observers restrict the objectivity of this test. It does, however, enable a single observer to compare the general optical performance of different binoculars with a high level of confidence. A table of appropriate double stars is given in Appendix 1.

- **Limiting Magnitude.** The standard way of establishing the limiting magnitude of an instrument is to count the stars in a known region of sky. Some useful regions applicable to binoculars are detailed in Appendix 2.
- **True Field of View**. When you are trying to find objects by star-hopping, it is essential to know what the true field of view of your binoculars is. The field of view that manufacturers state for their binoculars is not always correct, and, when wrong, they tend to err on the optimistic side. By placing stars of known separation at diametrically opposite sides of the binocular field, you can easily determine its true field of view. Similarly, if the field of view is severely degraded towards the periphery, you can determine the size of what you consider to be the usable field. A table of convenient star pairings, with relevant charts, is given in Appendix 3.
- **Field of View.** Field of view is a very personal thing. Some people seem to prefer a wider field of view, such as that from an ultrawide eyepiece giving 82° or so of apparent field of view, and perceive apparent fields of view of less than about 65° to be akin to tunnel vision. Others find that a narrower apparent field of view helps them concentrate on the object under observation and dislike having to "look around" to find the edges of the field of view.
- **Collimation.** Poor collimation is more obtrusive, and thus easier to test for, at night. There are several ways you can do this and most people will find that one way works better than others for them:

  1. Focus the binoculars on a reasonably bright star and then carefully move the binoculars away, keeping the image in view, until the binoculars are about 15–20 cm (6–8 in.) from your eyes. If you still have a single image, the binoculars are probably collimated within acceptable limits. If the images from each optical tube separate from each other, then the binoculars are miscollimated and will cause eyestrain.
  2. Defocus one side of the binocular so that a bright star becomes a large blur. Observe a bright star through both tubes. The focused image should fall exactly in the center of the defocused image. If so, the binoculars are probably reasonably well collimated.
  3. Use the red/cyan anaglyph method mentioned above.
  4. Use the *crossed Bahtinov* method devised by Konstantinos Makropoulos. This requires two Bahtinov masks, one on each objective, orientated at 90° to each other (Fig. 4.5). The masks enable a bright star or planet to be focused very precisely. If the binoculars are collimated, the centers of the diffraction flares

Field Tests

Fig. 4.5 Crossed Bahtinov masks

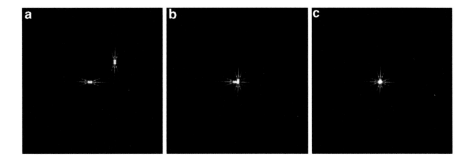

Fig. 4.6 (**a**) Badly collimated. (**b**) Nearly collimated (probably acceptable). (**c**) Perfect collimation

coincide (Fig. 4.6). This method is very sensitive and, performed with care, can detect miscollimation that is well within industry-standard limits. For the purposes of this test, the masks can be made cheaply and simply by printing them out on transparency sheet. There is a web page that will generate an appropriate Bahtinov mask at http://astrojargon.net/MaskGenerator.aspx. An assumed focal ratio of f/4 will generate suitable masks for most binoculars.

You should not expect to find a binocular that is perfect in every respect, but these simple tests ought to enable you to make a reasonably good assessment of the binocular in hand and to compare it to other binoculars.

Additional Tests for Used Binoculars

Used binoculars have potential deficiencies that are unlikely to exist in new instruments.

Give the binocular a thorough visual inspection for **evidence of repair or tampering**. If there is any, try to find out what has been done.

Wherever possible, check for damage **under eyepiece rubbers** (by touch, if the rubbers cannot easily be lifted). Damage to the eyepieces and their surrounds are the most common form of damage to binoculars.

Inspect external **optical surfaces for scratches**. This is best done viewing the surface at a glancing angle, which may reveal fine scuffs that result from improper cleaning. Eyepieces are particularly prone to this, since they tend to accumulate more debris, which is often wiped off with the first bit of clean-ish cloth that comes to hand, e.g., a tee-shirt hem or similar.

Look for evidence of **failing optical adhesive** between lens elements. This may have a milky appearance in patches and often starts at the edge of the lens. It is usually prohibitively expensive to correct.

Look for signs of **fungal growth**. This is most likely to be seen on the edges of lenses, where it can penetrate between the lens elements. If it has penetrated, it is expensive to correct.

Compare the view through both sides. If there is **internal optical damage**, it is usually not symmetrical, so one side of the binocular may have a different apparent color, a different clarity, or a different amount of light scatter.

Chapter 5

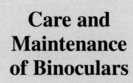

Care and Maintenance of Binoculars

Binoculars are generally robust and require very little maintenance if they are properly cared for. The exceptions are some of the budget binoculars that have flooded the market over the last couple of decades, many of which seem to have poorly mounted prisms that are easily knocked out of alignment.

As with any other optical equipment, indeed *any* equipment, prevention and preemptive maintenance are considerably preferable to curative maintenance and repair. There are five categories of foreign matter that threaten the well-being of binoculars. In no particular order these are:

- **Moisture.** This can invade the binocular either by condensation or from direct exposure to water.
- **User-originated grime**. This includes grease from fingers and eyelashes, spillages of food and drink, hair, and flakes of dead skin.
- **Environmental grime.** Dust is the usual culprit here, but grit can also enter the binocular in some circumstances.
- **Flora**. The usual culprits are algae and fungi.
- **Fauna**. Arthropods, especially insects and arachnids, can find ideal homes in the nooks and crannies of binoculars and their cases. William Gascoyne invented the eyepiece reticle after a spider had spun its web near the focal plane of his telescope!

Rain Guards

A rain guard is a cover for the eyepieces of a binocular. It is intended to protect the eyepieces of binoculars used for birding, etc., when they are slung from the neck in the rain or snow. As observational astronomers, we tend not to pursue our hobby in the rain, so precipitation is a relatively unlikely source of moisture. Despite this, a rain guard is an invaluable addition to any binocular, particularly one that is hung from the neck. Rain guards protect the eyepieces against spillages of food and drink and against any other descending particulate matter, whatever its origin. For this reason, a rain guard is most valuable if it is of the kind that can be attached to the neck strap so that it is immediately available for use. It soon becomes second nature to put the guard over the eyepieces when you pause in observing and the reduction in frequency with which eyepieces need to be cleaned is very noticeable (Fig. 5.1).

Storage

Although moisture does not usually affect an astronomical binocular from precipitation, it can do so from condensation. In terms of frequency of occurrence of damage, condensation is by far the most harmful source of deterioration of astronomical binoculars. Condensation on the external optical surfaces results in their being wiped more frequently than would otherwise be necessary, with the attendant damage that all too often accompanies frequent cleaning. (To protect against condensation in use, see the section on dew prevention in Chap. 8.) Condensation on internal

Fig. 5.1 Eyepiece rain guard. The rain guard attaches to the strap and can be slipped over the eyepieces when the binocular is not being used

optical surfaces may lead to inexpert dismantling of the binoculars, which itself can lead to damage, in an attempt to gain access to and clean the affected surfaces. The water itself accelerates the corrosion of the metal parts of the binocular, especially if there are places where two different metals are in contact. This corrosion leads to stiffness in moving parts and this stiffness, allied to the corrosion, accelerates the wear of these parts. Lastly, moist surfaces are a sine qua non for the growth of algae and fungi; if you keep moisture at bay, you will keep these flora at bay.

Much of this condensation damage ultimately results from the practice of failing to cap the binoculars at the end of an observing session. The cold binocular is taken into a relatively warm and humid dwelling, where moisture from the air condenses on the colder surfaces of the binocular. Lens caps and cases then act to hold the moisture in place. It is good practice, if you bring the binoculars indoors uncapped, to place them horizontally on a firm flat surface in the room in which they are to be stored until they have reached thermal equilibrium, then cap them and put them in their cases. It is even better practice to cap them out of doors before bringing them in *unless the optical surfaces are affected by dew*.

Ideally, binoculars should be stored in a cool dry place; certainly not one that is subject to great fluctuations of temperature and humidity. I store mine in a closet in an unheated part of the house. Cases and caps help to guard against arthropods that might otherwise take up residence. Multipurpose "grab-and-go" binoculars are often stored, uncased, and uncapped, on interior windowsills, where they are instantly available should an object of interest be spotted. They are placed resting on their objective ends, so the objective lenses are protected to some extent. On the other hand, the eyepieces of such binoculars act as dust magnets and soon show the marks of frequent scouring. If you must store your multipurpose binoculars in this manner, at least use eyepiece lens caps or a rain guard to keep the dust at bay. Even better, store them somewhere to hand where they are less likely to be dislodged by other household members and are less likely to have to endure the tremendous temperature ranges to which a black object on a sunny windowsill is subjected. Those who store binoculars on window sills are second only to those who use unmounted binoculars without a neck strap in ensuring the survival of the binocular repair industry!

Desiccants

Some years ago I used to maintain a 20×120 naval binocular of 1940s vintage. It had two small mesh-lined inserts into which a desiccant, presumably anhydrous calcium chloride, could be placed. This would reduce the likelihood of condensation on the internal optical surfaces. Nowadays, we do not use internal desiccants but rather fill the binocular with dry nitrogen and make it gastight. Nevertheless, desiccants still have their place. Nowadays, the preferred desiccant is silica gel, an amorphous form of silicon dioxide that can absorb up to a third of its own weight of water onto its surface (of which it has about 700 m^2 per gram!). It can be regenerated by heating it in an oven to between 125 °C and 200 °C (250° and 400 °F).

Sachets of silica gel are included with a multitude of modern electronic devices—and with binoculars. I tape a sachet to the inside of each of my small binocular cases and also to the inside of each objective lens cap of my 100-mm binoculars. This is possibly a bit of "overkill," but it seems to be little effort to eliminate an inconvenience that may necessitate a far greater effort to remedy.

Grit

If binoculars are placed in contact with a gritty surface, it is all but inevitable that some grit will end up where it is not wanted. Sea sand is the most pernicious species of grit; even if you ensure that your hands, and anything else that touches the binocular, are meticulously clean, you can almost guarantee that windblown sand will find a way in. The result is that telltale crunching sound as the abrasion of a moving part starts to occur!

In reality, you cannot expect to keep grit entirely away from the binoculars if they are actually going to be used. You can take the obvious precautions (above) to reduce the contact with grit and also try to ensure that no grease seeps from the moving parts of the binocular. Grease traps grit. It also tends to get transferred, with or without the grit for which it is often a vector, to optical surfaces. Particularly unpleasant is the grease that is used in some binoculars of Far Eastern origin; it seems to be more akin to an adhesive than to a lubricant. It is important, therefore, to remove any exposed grease. This is most easily done with a paper towel or tissue, since paper tends to absorb oils and grease.

Cleaning

It cannot be overemphasized that by far the best form of cleaning of optical surfaces is to prevent the dirt from accumulating there in the first place, i.e., use rain guards, lens caps, and cases whenever appropriate. The reason for this is that it is extremely easy to damage optical surfaces by cleaning them. In reality, an optical surface has to be quite filthy before the dirt has an optical effect that is significantly noticeable during visual observation and there is a real need to clean it. The objective lenses of my binoculars can go for years without being cleaned, and the eyepieces may only be cleaned once a year, although those that I use for star parties tend to need a clean after each event. In general, eyepieces can be cleaned more often than objectives. The lenses are smaller and can therefore be more easily hard-coated without introducing thermal stress into the lens.

My full binocular optical cleaning kit (Fig. 5.2) consists of the following items:

- **Puffer brush with retractable soft bristles**. This is my first line of attack. Most dust, etc., can be blown off with the puffer alone. If it is more stubborn, I deploy the bristles and use them in conjunction with the puffer. Flick the brush quickly over the lens surface while "puffing" the bulb. After each stroke across the lens, flick any accumulated dust off the brush, while giving a sharp puff to help

Fig. 5.2 Cleaning kit. *Top* L-R: lens tissue, microfiber cloth. *Middle*: camel-hair brush, blower brush, *Opti-Clean®*. *Bottom*: *Lens Pen®*

dislodge it. Be very careful if you use canned air as a blower—the propellant can damage lens coatings.
- **Microfiber cloth**. I keep one of these in each binocular case. To use it, hold the binoculars so that the affected lens surface is facing down, then gently flick an edge of the cloth over the lens to remove any dust. If any deposit remains, breathe on the lens to moisten it slightly, then *gently* wipe the lens from center to periphery. Microfiber is quite good at removing grease and finger prints. Never rub the lens with a circular motion and never wipe from the periphery to the center. If there are trapped particles that could scratch the optical surface, they will be trapped in the lens surround; they are relatively safe there, so don't accidentally dislodge them.
- **Lens tissue**. Lens tissue is particularly useful on small lenses. It can be made into a swab rather like a cotton bud by folding it into a strip and wrapping it around the end of a toothpick or similar. Dampen it in cleaning fluid and use it in light strokes from the center to the periphery of the lens.
- **Lens Pen™**. A *Lens Pen®* incorporates a retractable soft brush and a cleaning pad that is recharged with dry polishing agent from an impregnated piece of foam in its cap. It is particularly good at removing eyelash grease and fingerprints. First of all, use the brush to remove any dust. When you are sure that the lens surface is clear of particles, breathe on it to moisten it and clean the deposits with the pad, taking care not to drag any particles from the lens surround. I only use this on the hardened coatings of eyepieces and only when "in the field" where I don't have my liquid cleaners. It leaves an almost imperceptible thin film on the lens surface; this can later be removed with a liquid cleaner.
- **Opti-Clean™ and First Contact™**. *Opti-Clean™* and *First Contact™* are polymer-based cleaning product that were developed to clean silicon wafers for

use in the microelectronics industry and which are now marketed primarily for optical surfaces used in holography. They are suitable for any glass lens. They come in the form of a transparent liquid that is applied to the lens surfaces. As the liquid dries, it forms a skin. When the skin is peeled off with an adhesive tab, it pulls any grease and grime with it, leaving the lens in a pristine condition. It seems expensive, but it lasts for a very long time—I have had a 5-ml vial of it for about 20 years. *First Contact®* also comes with the option of having a red dye added to it. In this form it is intended for protected storage: the red dye immediately indicates that the optic is coated.

N.B. There are different products with the same name used for contact lenses and in dentistry; these are not suitable for astronomical optics.

- **Optical Wonder Fluid™** is a proprietary antibacterial lens-cleaning fluid produced by *Baader Planetarium®*. It should be applied to a fine microfiber cloth, which is then used to clean the lens. Never apply this sort of cleaning fluid to the lens itself, as it will seep down the sides. There are other proprietary cleaning fluids available from photographic outlets. Alternatively, you can make your own. My recipe is:

    Six parts distilled water
    One part pure isopropyl alcohol (IPA)
    Two drops liquid detergent (e.g., mild washing-up liquid)

Apply it to the lens with a lint-free cotton swab or a swab made of lens tissue wrapped around the end of a toothpick. Dampen the swab and swab the lens from center to periphery, rolling the swab so as to lift any grime away from the surface. Be careful not to over-moisten the swab otherwise you may get liquid into the lens surround. You can dry the lens with a dry swab or with a clean lens tissue.

However you clean the lens, be careful not to rub it any more than absolutely necessary. Not only does rubbing increase the likelihood of damage to the lens surface or coatings, but rubbing with a dry cloth or tissue can cause the buildup of a static electric charge on the lens surface. This charge will attract dust or lint, and it will be extremely difficult to dislodge it. If this does happen, you will need to use a water-based cleaning solution (such as the one above) to get rid of the static electric charge.

Dismantling Binoculars

In general, you should not attempt to dismantle your binocular. Unless you know precisely what you are doing, you run the risk of causing more damage than you are attempting to remedy! You would also immediately void any warranty. The single exception for the lay person is when a faulty binocular has been pronounced, by someone who is qualified to do so, unworthy of repair, either due to the extent of the damage or because the cost of repair would exceed the value of the binoculars. In these instances, there is a valuable experience to be gained.

Fig. 5.3 Objective barrel unscrewed. The prisms are visible under the cover plate

The faults which can be relatively easily repaired are dust or flora in the binocular, prisms that have been displaced by impact, and grit or failed lubrication in moving parts. The simplest binoculars to dismantle are Porro-prism varieties. If you attempt this, which you do entirely at your own risk, first prepare your work surface. I prefer to cover the work surface with plain white paper—I have sheets of acid-free tissue which is ideal. I have a few plastic containers for components, and I ensure that any tools I use are clean. I use disposable powder-free latex gloves to handle optical components.

If the binocular has the Zeiss type of body, the objective tubes can be simply unscrewed from the binocular body (Fig. 5.3), giving access to the inside surface of the objective lenses. The lower prism cover plates can then be removed, giving access to the lower prisms. These are often held in place by clips. In the Bausch & Lomb type of body, the objective cell must be removed from the tube. First of all remove the front protective ring—this usually simply unscrews. In this type of binocular housing, you do not gain access to the prisms from this end.

Fig. 5.4 Tripod bush removed. This gives access to the screw that secures the focus shaft

The objective lenses are held into their cells (which, in Zeiss body types may be integral with the objective tubes) by a locking ring and usually a seal ring. The locking ring should be removed using a peg spanner, but this can be done **with extreme care** with a small screwdriver whose blade fits into a recess on the ring. The danger of scratching the lens is very great indeed, and you should **not** attempt this unless you are willing to accept this risk. Before removing the locking ring, mark the eccentric rings with a soft pencil in order that they can be returned in the same orientation. If you remove the lens elements, mark the edges with a soft pencil with an arrow pointing to the front surface. Wrap them in acid-free tissue and put them safely in a container.

The dismantling of the eyepiece end begins with the removal of the cover at the bottom of the hinge. If there is a tripod bush in the hinge, you will also need to remove this; it is slotted for a screwdriver for this purpose (Fig. 5.4). This reveals a hole in the hinge, deep within which is the screw that holds the focus shaft in place. Use a flashlight to ascertain what type of head the screw has, and insert a screwdriver into this hole and undo the screw. A small amount of *Blu-Tack®* or other similar adhesive putty on the end of the screwdriver aids the removal (and replacement) of this screw (Fig. 5.5). When the screw is removed, use the focus wheel to drive out the eyepieces and bridge (Fig. 5.6). The focus shaft should be greasy; do not allow this grease to get onto optical surfaces or parts that may transfer it to optical surfaces.

Once the eyepieces and bridge assembly is removed, the eyepiece guide tubes must be unscrewed (Fig. 5.7). Then remove the screws that secure the top cover plate in place and remove the cover plate (Fig. 5.8). Ensure that you store screws in such a way that you know which screw goes where. In Zeiss-type bodies, the

Fig. 5.5 Adhesive putty holds the screw to the screwdriver

Fig. 5.6 The bridge and eyepieces are lifted clear

5 Care and Maintenance of Binoculars

Fig. 5.7 The eyepiece guide tubes are removed

Fig. 5.8 Undo the cover-plate screws

Dismantling Binoculars

Fig. 5.9 Clamped prism in Zeiss-type binoculars

Fig. 5.10 Prism cluster in situ

upper prism will be held in place by a clamp. This may be screwed down at one end (Fig. 5.9) or, in cheaper binoculars, have both ends clipped under recesses. The prism may also be protected by a shaped piece of metal or card-type composite.

In Bausch & Lomb type bodies, the prisms normally remove as a cluster (Figs. 5.10 and 5.11). The screws that secure it in place are the ones immediately adjacent to the slotted-head grub screws (set screws) that are used for collimation. The prisms are secured to the cluster plate with clamps that are screwed to the plate.

Fig. 5.11 Prism cluster removed

The eyepiece lenses are retained with a lock ring. If you decide to dismantle the eyepieces in order to clean the components, be sure to mark the edges of the various lenses and spaces so that you know their order of reassembly and the direction they should face. Also be aware that there may be as many as six separate lens elements.

The only time when it is necessary to dismantle the hinge is when the tension needs to be adjusted. Remove the cap with the IPD scale on it, and you will see a slotted brass tension screw with a locking grub screw (set screw) in it. If you need to adjust the tension, loosen the locking screw and adjust the tension screw with a screwdriver. The correct tension is achieved when it is just sufficient to prevent one side of the binocular sagging under the effect of gravity when the binocular is held by the other side.

When you reassemble the binocular, it may be necessary to lubricate some of the mechanical parts. Use a good-quality lithium grease for this. Use the minimum amount necessary and ensure that none is able to escape to the outside, where it will inevitably be transferred to the external optics. If screw threads are slightly stiff and are tending to bind, you can lubricate them by running with a soft graphite ("lead") pencil along the thread. (Incidentally, soft pencils are also useful for lubricating stuck zips (zippers) and stiff lock mechanisms.)

Right Eyepiece Diopter Adjustment

It is sometimes the case that the right eyepiece diopter adjustment of budget binoculars is poorly set. People whose eyes are similar often need to adjust the right eyepiece close to the end of its range in one direction or the other, when they would have expected it to be approximately central. This situation is worse with a user whose eyes are different by a diopter or more; in this circumstance, a badly adjusted right eyepiece may prevent the user from being able to focus both eyes.

The remedies can be simple:

- If the binocular is still under warranty, return it to the vendor.
- If you want to have a go at fixing it yourself, read on.

The Solution

The only tool you need is a small flathead screwdriver.

1. Carefully pry off the rubber eyecup; it may be held in place by a few smears of contact adhesive, but there is no need to reglue it when you replace it. You will probably uncover some of the adhesive substance that the manufacturer has substituted for a decent lubricant grease—do your best not to transfer any onto the optics.
2. Locate and slacken the three set screws (aka "grub screws") that are located under the flange (Fig. 5.12).

Fig. 5.12 Locate the set screws (*grub screws*)

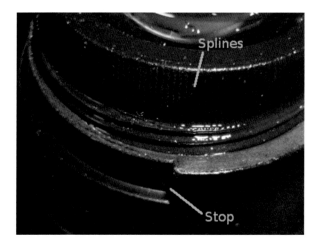

Fig. 5.13 The rotating part has been removed

3. Rotate the eyepiece as far as it will go in the direction you need to extend the rotation.
4. Lift off the outer rotating part to reveal the adjustment mechanism. Note the splines on the rotating lens housing and the stops on the fixed part (Fig. 5.13).

Warning: In some examples of this binocular, the outer rotating part is cemented to the inner rotating part and will not easily lift off. If it is not cemented, removal takes no more force than it does to rotate the mechanism.

If it is cemented, it can still be removed with a lot of force and extreme care but *you will probably damage the binocular.* If it is cemented, there will be no splines and you will need to re-cement it once you have adjusted it to your liking. Take care not to get adhesive onto the eyepiece or into the threads of the adjustment mechanism.

5. Note the lug and splines in the rotating part that you removed in the previous step (Fig. 5.14). Replace this part so that the lug is midway between the stops. Test to see if you can obtain focus with both eyes. If you cannot, repeat steps 3–5, and obtain the correct adjustment by trial and improvement.
6. Once you are satisfied, reassemble the eyepiece in the reverse order to disassembly. Remember to tighten, but do not overtighten, the grub screws. Replace the rubber eyecup so that the central diopter mark aligns with the fiducial mark when the right eyepiece is correctly focused for you.

Fig. 5.14 The lug that prevents over-rotation

Collimation

If miscollimated binoculars are still under warranty, return them to the supplier. The supplier should have access to a binocular repair shop which has proper collimating equipment. Proper collimation is a skilled task and can be expensive on binoculars are out of warranty. It can cost more than the binoculars cost in the first place and is therefore usually not worth having done on budget-priced binoculars if they are out of warranty. If your binoculars are out of warranty and you feel confident of trying to do it yourself, here is how. The methods described in this book will result in conditional alignment,[1] i.e., the optical tubes will only be aligned at the interpupillary distance at which you perform the alignment. Binoculars can be collimated either by eccentric rings on the objective lenses or by tilting the prisms with grub screws (set screws). There is no substitute for experience in collimation. If you can, practice on an old misaligned binocular where you will not be upset if you are unable to achieve the collimation you want.

Always collimate binoculars out of doors or indoors by looking through an open window. Window glass is usually nonuniform and can differentially affect what you see through each side of the binocular. Collimate by looking at an object at least 500 m (550 yds) away. If the object is too close, its image will appear even closer, and your eyes will naturally converge while looking at it.

---

[1]"Conditional alignment" is a term introduced by the binocular repairman William J Cook. It describes the situation where the optical axes of the binocular tubes are aligned with each other at a particular interpupillary distance (IPD). They are not aligned to the binocular hinge. If the IPD is changed, the optical axes of the tubes will no longer be aligned, i.e., the alignment is conditional upon the IPD remaining unchanged.

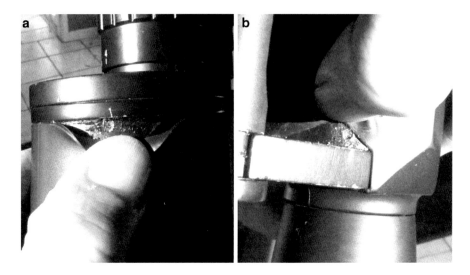

Fig. 5.15 The collimation screws may be under the rubber covering

If the binoculars were once properly collimated and have suddenly lost collimation, this is most likely due their having been dropped or a "bump" causing a prism to shift. It is therefore worth examining the prisms to see if there is any obvious shift. Often, if a single prism has shifted, a symptom will be that the image in the affected tube will have acquired "lean," i.e., it will be tilted slightly to one side or another. Prisms are held either to the body of the binocular (*Zeiss*® or "European" style) or in prism housings (*Bausch & Lomb*® or "American" style) by straps. A sharp jolt can move the prism and, if this has happened, it can usually be replaced. In budget binoculars, there is often no possible adjustment of the prism once it is located and secured into its recess in the binocular body (see Fig. 5.9).

Prism adjustment. If prisms are adjustable, this will either be by external grub screws (set screws) which are accessible under the rubber or leatherette covering of the binocular body, usually close to the edge of the prism housing where they can be accessed by minimal shifting and stretching of the rubber. The holes are usually filled with a rubbery adhesive substance that needs to be prized out with a small screwdriver to allow access to the collimation screws. If the binocular is covered with leatherette, there are usually little tabs in the leatherette that can be lifted to permit access to collimation screws. Alternatively, the collimation screws may only be accessible by removing the cover plate. They are the slotted-head screws (Figs. 5.10, 5.15, and 5.16).

Remove the eyepieces and bridge, then remove the cover plate and replace the eyepieces and bridge. The collimating screws are slotted grub screws that are immediately adjacent to the crossheaded screws that secure the prism cluster to the binocular body. The screws adjust in "push-pull" pairs. The procedure is to slacken

Collimation

Fig. 5.16 Collimation screws in a prism cage

the securing screw, adjust the collimating screw, then retighten the securing screw. Adjust one pair of screws at a time and turn the collimating screw no more than an eighth of a turn and examine its effect on the image. Adjust in small increments until the images are satisfactorily aligned.

Make sure that you do not inadvertently rotate a prism. Doing so will cause the image on that side to "lean." The angle of lean is in the opposite sense and twice the magnitude of the angle that the prism is rotated (see Fig. 2.3).

Eccentric rings. Binoculars differ from telescopes in that collimation is achieved by lateral movement of the objectives, not by tilting them. Moving the lens one way will move the image in the eyepiece the other way. However, it is all but impossible to move an objective only either laterally or vertically, and collimation with eccentric rings can be monumentally frustrating until you get the hang of it. It can be a tough test of perseverance and patience—you have been warned!

First of all, mark the positions of the rings (see Fig. 5.17) so that, if you do not manage to improve matters, you can at least set them to their original position. Next, set the rings so that there is no eccentricity, i.e., the narrowest part of the inner ring aligns with the widest part of the outer ring and vice versa. Rotate the objective lens assembly a small increment—say about 10–15°—at a time until it has made one revolution, and see whether there is any movement of the image and, if so, if it is sufficient to bring the images into proper alignment. If it is not, slip the inner ring about 10° and repeat the rotation of the lens assembly. Repeat this until the images are aligned as well as possible. If there is not sufficient eccentricity in the rings, you will need to adjust the prisms.

5 Care and Maintenance of Binoculars

Fig. 5.17 Eccentric rings. Rotating these rings moves the lens laterally. Note that it is marked to enable its return to the original position

Note: It is usually better to set the binoculars so that there is a discernible amount of convergence, i.e., so the optical paths from the binoculars are diverging (see Chap. 2, footnote 3), and then to gradually collimate from there than it is to approach collimation from the other way. This is because the eyes are more sensitive to divergence than to convergence. You may find them more comfortable if there is slight convergence (this is also known as the eyepieces being "coned in") because the experience is that the image we see is perceived by some people to be relatively close, at a distance at which their eyes would naturally converge.

For more detailed accounts of collimation see:

- J.W. Seyfried's *Choosing, Using, & Repairing Binoculars,* which gives a more detailed account of conditional collimation, including building bench-testing apparatus.
- The Naval Education and Training Program Development Centre's (NAVEDTRA) *Basic Optics and Optical Instruments*, which gives accounts of full collimation with bench test apparatus.

Bibliography

Dismantled Porro-prism binocular: http://www.actionoptics.co.uk/disdbin.htm
The Naval Education and Training Program Development Centre, **Basic Optics and Optical Instruments**, New York, Dover, 1997, ISBN 0-486-2291-8.
Seyfried, J.W., **Choosing, Using, & Repairing Binoculars**, Ann Arbor, University Optics Inc., 1995, ISBN 0934639019.

Chapter 6

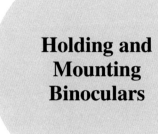

Holding and Mounting Binoculars

When you decide how to mount your binoculars, there are two considerations that you must take into account. These are stability and comfort. Both of these play a significant part in determining how much you will be able to see. If the binoculars are not held in a manner that is reasonably stable, in order to eliminate shake, the amount of detail that you will be able to see will be severely reduced. If you are not comfortable when you observe, you will quickly tire and tiredness is always detrimental to observing.

Hand-Holding

You will not see as much with your binoculars when you handhold them as you will if you mount them properly. However, if you use your binoculars for quick "grab-and-go" observing sessions, if you are using them for quick sky scans in conjunction with a telescope, or if you are using binoculars *because* they can be handheld and are therefore the ultimate in a portable observing instrument, then you will not want to use a mount. Small and medium binoculars up to about 10×50, depending on the ergonomics of the particular binocular, can be effectively handheld for medium periods. Slightly larger ones can be handheld for short periods and for "quick peeks." These periods can be extended and made more effective if ergonomic considerations are taken into account. Some people consider an extended discussion of hand-holding to be overkill, but if, by understanding and applying a few simple principles, you can increase the efficacy and enjoyability of your

Fig. 6.1 The "normal" hold

observing sessions, this discussion is worthwhile. There are four "basic" ways in which a binocular can be handheld:

The "Normal" Hold. It seems to be instinctive for most people to hold Porro-prism binoculars around the prism housings (Fig. 6.1). The weight of the binocular is taken entirely by the arms, and the upper arms are extended forwards. Although this initially feels comfortable, it is inherently tiring and unstable, especially when you are observing objects at high altitude. It rapidly results in fatigue which, even after a few minutes, is easily noticeable, both in the limbs themselves and in the amount that you can see. It is very easy to improve upon this, merely by moving your hands an inch or so closer to your face to form the "Triangular Arm Brace."

The Triangular Arm Brace. Many adults consider that 10×50 binoculars have too much magnification to be handheld. For several decades I have run astronomy clubs for youngsters and we have used 10×50 binoculars as our "standard" instrument. By teaching this way of holding binoculars, I have enabled children as young as 10 years old to effectively use these binoculars for observing. If the detail that they report as being able to see is any indication, they are seeing noticeably more than adults using the "normal hold."

Hold the binocular with your first two fingers around the eyepieces and the other two fingers around the prism housing. Then raise the binocular to your eyes and rest

Fig. 6.2 The triangular arm brace

the first knuckle of your thumbs into the indentations on the outside of your eye sockets, so that your hands are held as if you were shielding your eyes from light from the side. Rest the top knuckle of your thumb against the indent in the bone at the outside of your eye socket and the second joint against your cheekbone (Fig. 6.2). Each of your arms is now locked into a stable triangle with your head, neck, and shoulder as the third "side," thus giving you a much more stable support for your binoculars. Some of the weight is effectively transferred from your arms to your head and neck, so it is less tiring on the arms. The position of your thumbs keeps the eyepieces a fixed distance from your eyes. You cannot normally reach the focus wheel on center-focus binoculars when you hold them this way (although you can with roof prisms), but you should not need to refocus during an observing session. This grip does feel unusual at first, but it is so superior to the "normal" way that it soon becomes second nature.

The "Rifle Sling". If you want the most stability you can get with medium-sized handheld binoculars, use the "rifle sling" method. This is based on the way one uses a sling for rifle-range shooting, where stability of the rifle is improved if the arm is braced through the sling. Extend the binocular strap to its full extent and hold the binocular so that the strap loops down. Place both arms through the strap, so that it comes just above your elbows. Hold the binocular in the most comfortable way you

Fig. 6.3 The "rifle sling" hold

can and brace it "solid" by pushing your elbows apart (Fig. 6.3). It initially feels a bit like getting into some sort of medieval torture instrument, but it is remarkably effective for stability. Because it is not inherently comfortable, I do not use this method for long periods, but only for observations where I want that little bit of extra stability.

Double-Handed Hold. You may sometimes wish to handhold a larger binocular for short periods and find that the balance of the binocular makes the "triangular arm brace" method unstable. The double-handed hold uses both hands on one optical tube of the binocular. Assuming your right eye is dominant, use the "triangular arm brace" with your right hand and hold the right objective barrel of the binocular with your left hand (Fig. 6.4). You support the left objective barrel with your left wrist—if you wear a wristwatch or bracelets, you will probably find it more comfortable if you remove them first. For extra stability, you can combine this with the "rifle sling" method.

"Informal" Supports

Fig. 6.4 The double-handed hold

An inherent source of fatigue and instability with all hand-holding methods is that your upper arms are extended in front of you. If you can support your elbows and/or upper arms, you will notice a drastic improvement in stability and reduction in fatigue.

"Informal" Supports

The stability of a binocular can be markedly increased if you can somehow support your arms, most effectively done at the elbows. You can do this by resting your elbows on walls, automobile roofs, walls or fences, gates, boulders, rotary washing lines, tables, cushions, or other elevated supports on the arm rests of recliners, the shoulders of a companion, etc. This list is limited only by imagination and ingenuity. Even if you cannot support your elbows, merely leaning your body against a supportive object (wall, tree, automobile, boulder, telegraph pole—or the ground!) will confer increased stability to what you see in the binocular eyepiece.

Mounting Brackets

The most commonly used binocular mount is the photo tripod and L-bracket. Most modern binoculars in the aperture range 40–70 mm have, in the distal end of the center-post hinge, a bush for an L-bracket. Porro-prism binoculars in the aperture range 80–100 mm may either have a bush for an L-bracket or a bush for direct mounting on a tripod plate. Those of greater aperture than 100 mm almost always have a direct-mounting bush.

Some older 50-mm binoculars also have direct-mounting bushes, usually on the right-hand prism housing (Fig. 6.6). It has been suggested that this arrangement can cause the optical tubes to lose alignment with each other, but I have not found this to be the case. However, if you have this type of binocular, you do need to ensure that the tripod head allows it to be mounted in such a manner that your nose does not foul the mounting plate! A tripod bush like this offers no facility for tilting the binocular side to side. Some people deem this to be an advantage.

Brackets for mounting a binocular to a tripod come in four distinct categories (Fig. 6.5). It is important to acquire the one that is specific to your needs.

Hinge Clamp. As its name suggests, this clamps onto the center-post hinge of Porro-prism binoculars. It is used when there is no mounting bush on the binocular.

Fig. 6.5 Various mounting brackets. *Top, L-R*: Hinge clamp, universal L-bracket, roof-prism L-bracket, proprietary (Universal Astronomics) L-bracket. *Bottom:* Bush in prism housing

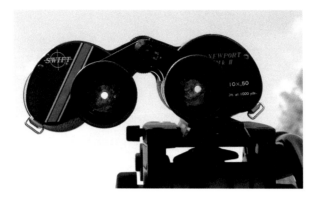

Fig. 6.6 Direct-mounting bush

Fig. 6.7 Hinge clamp

It may be unsuitable if the binocular has a wide-focusing "band" as opposed to a narrow-focusing wheel because, in the former, there is usually insufficient center post to accommodate the clamp. As most modern binoculars have a tripod bush, this will most often be necessary on older binoculars (Fig. 6.7).

Universal L-bracket. This fits almost all Porro-prism binoculars that have a ¼" mounting bush at the distal end of the central hinge. The bracket is usually merely a strip of metal that is bent into an L-shape, painted or coated, and furnished with the appropriate holes and bolts. Near the end of the upright of the "L", it has a captive screw that screws into the ¼" mounting bush of the binocular. The "foot" of

Fig. 6.8 Universal L-bracket for Porro-prism binoculars

Fig. 6.9 L-bracket for roof-prism binoculars

the "L" has one or two ¼″ threaded holes for the ¼″ screw on the tripod mounting plate. If there are two holes, use the one that offers the best balance to the binocular (Fig. 6.8).

Roof-Prism L-bracket. This is similar to the universal bracket but has recesses for the objective tubes of roof-prism binoculars. These are closer together than in Porro-prism binoculars and cannot be used with most universal L-brackets. Because the recesses have the potential to weaken the structure of the bracket, they are thicker from front to back than are universal brackets in order that they do not flex in use (Fig. 6.9).

Proprietary L-bracket. These work in exactly the same way as the universal bracket, but the "foot" of the "L" is adapted to fit a proprietary mounting.

Monopods

For several decades, photographers have been aware of the use of the monopod as an ultra-portable and compact camera support. It also makes an ultra-portable and compact support for binoculars. The mere fact that the binocular is supported confers a degree of stability that is not obtainable by hand-holding.

In order to use a monopod, you will need to obtain a suitable L-bracket or other mounting bracket for your binocular, unless it is already fitted with a mounting bush. These brackets are discussed in more detail in the section on tripod mounting, below.

Monopods can be used for both standing and seated observing, but few are sufficiently long for observing high altitude from a standing position and, in any case, it is extremely uncomfortable to try to observe at high elevations while you are standing, unless the binoculars have angled eyepieces. In recent years, it has been possible to purchase hiking poles in which the top part of the handle can be removed to reveal a camera-mounting screw.

Monopods can be made much more useful with the addition of a trigger-grip (aka "pistol grip") ball-head. As long as this can comfortably support the weight of the binocular, this is a very versatile combination and changes the monopod from being merely useful to being a pleasure to use. The setup shown in Figs. 6.10 and 6.11 is holding a 2.5 kg 15×70. If a monopod is not equipped with some form of moveable head, the entire monopod may have to be moved, and extended or collapsed, as you slew from one object to another. With a moveable head, this is enormously reduced; the trigger-grip makes it very much simpler, as you don't need to constantly adjust tension screws with cold or gloved fingers. The other hand can control monopod length, an action that very soon becomes intuitive. The monopod doesn't have to be vertical to take sufficient weight to relieve your arms and to steady the binocular. If you use a recliner with a fabric seat, you can keep it closer to vertical by making a hole in the seat between your legs, through which the monopod pole (and the small items that escape from your trouser pockets when you recline!) can pass.

This combination has become my favored mount for the 15×70 in the photograph, which has recently become my "grab-and-go" instrument of choice when I want to do some quick observing. The only steadier solutions for comfortable observing of high-altitude objects are parallelogram mounts and custom-built observing chairs, both of which detract from the portability of the binocular.

Fig. 6.10 Seated at monopod

Neckpod

An adaptation of the monopod is the "neckpod." This is a monopod that is suspended from a broad strap around the neck. The one shown is intended for small cameras, but is quite usable for the 1.3 kg 15×70 shown in the photograph. Although it is not as steady as a proper monopod, and is being used with a weight at least twice as great as what it was designed for, it is a noticeable improvement on hand-holding. Most of the weight of the binocular is taken by your torso, so it is far less tiring than hand-holding. However, it can be quite fiddly to use although, once you get the tension in the tilt head set to something usable, all you need to adjust is the monopod length when you change elevation (Fig. 6.12).

Note: The tilt head cannot be tightened enough to support the cantilevered weight of the binocular, but you don't want it to be immovable, or it makes it difficult to

Fig. 6.11 Trigger-grip ball-head

Fig. 6.12 Neckpod

slew to objects of different elevations. You need to set it to a tension that enables you to move the binocular, but not so loose that it swings freely if you remove your hand from it.

Bodge-o-pod

In extremis, a support can be bodged from household items. A simple but very effective "poor-person's" alternative to a monopod is a humble broom or mop (with a clean and dry business end or a clean and dry cloth over it!). You can easily secure the binoculars to the broom- or mophead with a bungee cord, although many people, myself included, prefer merely to rest the binocular on the head. If the mop has a telescopic pole, such as is found in several designs intended for window-cleaning, then its utility is increased as these usually extend to a length that is suitable for high-altitude observing from a standing position. They also allow relatively simple length adjustment when, in a seated position, you move from one object to one of a different altitude. A mophead with an adjustable angle is somewhat useful, but expect to have to overcome the same limitations as those that exist with the neckpod (Fig. 6.13).

Fig. 6.13 Bodge-o-pod

Photo Tripods

Photographic tripods, used with a "normal" photographic tripod head, are usually seen as the most obvious low-cost way of mounting binoculars. Unless the binoculars have angled eyepieces, photographic tripods and heads are not ideal observing platforms as it is extremely difficult to use them for observing at high elevations. Because photographic tripods are so widely used for their intended purpose, they can be mass-produced in great numbers and are readily available at a relatively low cost. However, all tripods are not equal and one that is suitable for binocular observing should meet several criteria.

- The tripod should be high enough to permit observations of object near the zenith while you are standing. If you try to observe near the zenith from a seated or reclining position using a tripod and normal photographic head, it is a near certainty that your legs and those of the tripod will, at some stage, need to occupy the same location; the consequences are, at best, infuriating. This requirement automatically eliminates the vast majority of photo tripods.
- The height of the tripod head needs to be adjustable so that, for a single observer, the height of the eyepieces of the binocular can be changed over a range of a minimum of 150 mm (6 in.) for straight-through binoculars and 250 mm (10 in.) for those with 45° angled eyepieces. However, these are not the total distance through which the tripod head must move. As the binoculars are angled upwards, the eyepieces get lower by an amount that is the sum of the distance from the eyepiece to the mounting bush and the distance from the mounting bush to the altitude bearing on the tripod head. To this must be added the difference in height between the tallest and shortest observer who will use this setup in a single observing session (but this latter requirement can be reduced if some sort of simple observing platform, such as an upturned milk crate, is available for shorter observers). This height adjustment requires some form of center post. The only usable types are those with a handle and ratchet for adjusting the height; those that work on friction alone are difficult to use for our purposes.
- The tripod head needs to enable observation near the zenith. Many of the more robust heads only enable, when they are used as intended, an elevation of about 60° (Fig. 6.14). However, many (but by no means all) of them allow a *depression* of 90°. If this is the case, it may be possible to use them reversed (Fig. 6.15). If this is the case, any handles will also need to be reversed and you should ensure that, in doing so, they do not obstruct or interfere with any locking or friction knobs that you will need to use when you are observing.
- The altitude bearing of the tripod head must be sufficiently robust, and have sufficient friction, to bear the turning moment of the binoculars when they are pointed near the zenith. The great majority of photographic heads are not designed to accommodate this sort of turning moment, which is rarely encountered using a consumer compact or DSLR camera, and are inadequate for the task. As a consequence, it is usually better to consider a robust video head with fluid motions. Test it to verify that it is sufficiently robust and has sufficient

Fig. 6.14 With the video head the right way around, the zenith is inaccessible

friction control to enable the binoculars to be pointed at the zenith *without changing aim when you release the handle*. All photo tripods that I have encountered that meet the previous criteria for the tripod itself have the facility to enable the heads to be interchanged, so this need not be a concern. Indeed, for most of these, the tripods and heads are sold separately.

Even if the tripod and mount do meet all the criteria above, they can still be difficult to use, especially when you are observing near the zenith with binoculars with straight-through eyepieces (unless you are an accomplished limbo dancer). The neck strain induced by using such a system for near-zenith observing from a standing position is extremely uncomfortable for most observers and very soon results in fatigue. The experience of keen kite flyers, who also spend long periods looking up into the sky, is that this posture can induce neck problems. The other shortcoming of a photographic tripod and head arrangement is that, sooner or later, your legs and those of the tripod (and, if you are seated, those of your chair) will all be vying with each other to occupy an identical bit of space-time. It's not usually an insurmountable problem—some rearrangement will usually suffice—but it is an

Fig. 6.15 Reversing the video head makes the zenith accessible

unnecessary irritant that can easily be eliminated. Hence, it is unsurprising that many of those who use tripod-and-head arrangements soon seek a different solution. This brings us into the realm of proprietary binocular mounts and observing chairs.

Fork Mounts

A solution to the cantilevering of heavy binoculars that are mounted on a photo or video head is the fork mount. It is possible to fork-mount binoculars so that they rotate in altitude about their center of mass; indeed, several large binoculars are provided with mounting bushes in the sides of the optical tubes for precisely this purpose. The yoke needs to be offset, so that the altitude fulcrum is not above the fork's azimuth axis, if it is to enable the binoculars to reach the zenith. However, this offset, and the consequent cantilevering, is constant, so it can be allowed for in the design.

If necessary, it can be compensated for by the use of a counterweight system. A fork mount does not eliminate the need to raise and lower the mount for observing at different elevations.

Mirror Mounts

Over the past few decades there have been various designs of binocular mount that use a first-surface mirror arrangement to circumvent the problem of an uncomfortable observing posture. These usually need to be placed on a table or tripod and the binocular is secured to the mount. Either the binocular and mirror or only the mirror itself can be rotated about a horizontal axis for altitude, and the entire mount may be provided with a lazy-Susan type, or other rotatable, base for azimuth adjustment. Such an arrangement is the popular *Sky Window*®[1] and is also amenable to DIY construction (Fig. 6.16).

These mounts are like *Marmite*®—nobody seems to be ambivalent about them and observers seem to divide strongly into two diametrically opposed camps: those who extol their virtues and those that loathe them; in the interests of enabling you to assess my objectivity, I have to declare myself to be a member of the latter camp, although I have only briefly used a mirror mount and have never used the aforementioned *Sky Window*®.

Fig. 6.16 A homemade mirror mount (Photo courtesy of Florian Boyd)

---

[1] http://www.tricomachine.com/skywindow/

Their advantages are obvious and simple: they permit observation, particularly of higher altitudes, from a normal seated position and with the head at a range of angles for which the human body seems to be naturally designed (the "microscope position"), thus eliminating neck and back strain and the resulting fatigue. The table, if one is used, also provides a rest for the elbows. There is no doubt that, from an ergonomic point of view, they are exceptionally comfortable and are an ideal solution for those observers for whom this is a major consideration. They are also compact and relatively light, so they are relatively portable. If tables and chairs are not available at the observing location, it is no greater hardship to carry a portable/ collapsible table and chair than it is to carry a tripod.

Their obvious disadvantages are that they provide an inverted image of the sky and that, unless they are provided with some sort of dew heater, the mirror is prone to dewing. Some observers find them difficult to aim. I have not seen any that are suitable for use with binoculars bigger than about 80-mm aperture. There are also disadvantages of using an additional optical surface. There will be some light loss although, as long as the mirror surface is of reasonably good quality, it will not be to the extent that it will be noticed in use by most observers. The mirror will also impose a limit to the amount of magnification used, due to the difficulties, and concomitant expense, of making a large optically flat surface. This is not usually a problem with magnifications of less than about ×15. Finally, there is the problem of cleaning. It is inevitable that such a large exposed surface will accumulate dust and debris; as with all first-surface mirrors, cleaning must be undertaken with extreme care so as not to damage the surface.

Parallelogram Mounts

Parallelogram mounts solve many of the problems inherent in the use of photographic tripods and heads:

- They move the observer away from the tripod so that its legs do not interfere with the observer's body position, especially when observing at high elevations.
- They offer easily changeable eyepiece height over a wide range and can thus accommodate different observing positions and observers of different heights. The eyepiece height can be changed without changing the aim of the binoculars, making them ideal for communal observing.
- The mounting head can be designed so that the binocular's center of mass can be aligned with the altitude fulcrum, thus eliminating problems associated with a changing turning moment when objects of different altitudes are observed.
- They are amenable to home construction by moderately competent wood- or metalworkers.

Their disadvantages are that they are relatively bulky, they require counterweights, and the long arms mean that vibrations take longer to damp down.

The simplest incarnations of the parallelogram mount have only altitude adjustment in the binocular mounting head, thus requiring that the observer moves in a circle around the tripod in order to change the azimuth. As more degrees of freedom

Fig. 6.17 A well-designed parallelogram mount allows more than a quarter of the sky to be observed without the observer having to move

of movement are introduced in the head, so more sky is observable from a single position. The paragon of this development is the *Universal Astronomics®*[2] deluxe mounting head, which enables more than a quarter of the sky to be observed from a single position (Fig. 6.17). A well-designed parallelogram mount, which has smooth motions and permits proper balancing of the weight of the binocular, almost confers the feeling that the binoculars are floating in the air in front of your eyes.

If you decide upon a parallelogram mount, you should give careful consideration to the length of the parallelogram arms. Longer arms enable a wider variety of observing postures, so that you can change from standing, through sitting, to reclining, without having to adjust the tripod height. Shorter arms have smaller

---

[2] http://www.universalastronomics.com/

damping times for vibrations but require that the tripod height is adjusted for different observing postures.

Parallelogram mounts, particularly those designed for big and giant binoculars, expose the limitations of photographic tripods, particularly those with center posts. The usual solution is to use a surveyor tripod. These usually do not have leg braces, but have spiked feet which press into the ground. This is ideal, and provides an exceptionally stable platform, if you observe on a surface where this is possible. On the other hand, if you observe on a surface that is unsuitable for this, you must acquire either spreaders or leg braces (which can be retrofitted to the tripod) or an expensive accident resulting from a slipping leg is all but inevitable. Do not be tempted to rely on being able to tighten the leg hinges sufficiently to prevent this. Spreaders can be obtained from most suppliers of survey tripods, and leg braces are available from *Universal Astronomics*.

Observing Chairs

Some form of observing chair can enhance the comfort of astronomical observing with binoculars, as well as being extremely useful for naked-eye observation and enjoyment of the heavens. There are a variety of options, ranging from simple inexpensive garden chairs or reclining loungers, through a plethora of dedicated homemade designs, to commercially available devices that are motorized in azimuth and altitude.

If you use binoculars because of their portability, an obvious choice is a collapsible recliner that is designed for portability. They are also suitable for use with a parallelogram mount (Fig. 6.18). These come in their own carrying bag, usually with a shoulder sling. Features to look for include sturdy construction, good comfortable support for your head and legs, and a continuous range of reclining positions that are easy to change by pressure of legs or shoulders, but which do not change involuntarily. It is not necessary for them to recline to a horizontal position; 30° to the horizontal is adequate. Almost all of these chairs are thoughtfully provided with cylindrical mesh accessory holders in the arms; you can use these for lens caps, for eyepieces (if your binocular has interchangeable eyepieces), for spectacles (if you remove them to observe), for pencils for recording observations, for a red-light torch (flashlight), or even for an insulted mug of warm drink or other refreshment. Some also have a pocket in the back of the seat that can act as convenient storage for observing charts or for this book. You may wish to add facilities to them. It is simple enough to suspend a fabric pocket for notebooks, etc., from an arm of the chair, and, if you are more comfortable with a cushion under your head, a fabric pocket for this can be attached in the appropriate place or you can merely hold a cushion in place with bulldog clips. A relative shortcoming of these collapsible recliners is that, unlike traditional garden loungers with wooden or rigid plastic arm rests, it is not simple to add extensions to raise the height of the armrests so that they can support your arms when you are hand-holding binoculars. However, a simple wooden frame could serve this purpose.

Fig. 6.18 *Mac sports* **recliner.** This well-designed folding recliner is ideal for use with handheld binoculars or with mounted binoculars. Note the high, padded headrest, which is a must for a recliner like this

Fig. 6.19 Homemade observing chair. Craig Simmons' chair features a rotating base and spring counterweighting for the binoculars. Note also how the altitude pivot axis coincides with the axis at the top of the observer's spine (Photo courtesy of Craig Simmons)

If you are a competent wood- or metalworker, you may be attracted by the idea of making your own observing chair. There is a multitude of designs of varying complexity published on the internet. Many of these are based on materials to which the constructor has easy access, and many constructors are somewhat unobjective in their evaluation of their own designs, so you do need to exercise some thought and care if you choose to copy one of these. A better option is often to adapt published designs to your own specific conditions of skills and availability of tools and materials. The URLs of some of the better designs are given at the end of this chapter (Fig. 6.19).

For the serious binocular observer, the epitome of commercial binocular observing chairs is probably the *Starchair*®.[3] This device is capable of supporting giant binoculars as large as 25×150, is fully motorized in azimuth and altitude, and has joystick control of the orientation of the chair. Although you may think that a device such as this warrants its own observatory housing, the *Starchair*® is fully portable in a small car! Although it comes with a hefty price tag, it is notable that it is the only commercial motorized binocular chair that has survived more than a few years in production (Fig. 6.20).

---

[3] http://www.starchair.com/

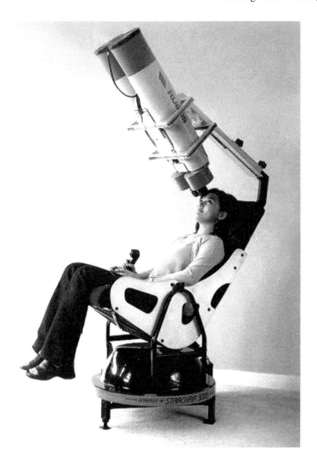

Fig. 6.20 *Starchair®*. Superb, commercially made, computerized observing chair (Photo courtesy of Starchair Engineering Pty Ltd)

Summary

There are several different issues involved with comfortable binocular viewing. In no particular order these are:

1. For objects above about 45°, if you don't want neck ache, you need something that enables you not to have to tilt your head back.
2. If you use a tripod without something that holds the binocular away from the tripod, sooner or later your legs and the tripod's will compete for the same space.
3. Whatever you use will need to have easy height (and, if you are seated, lateral) adjustment unless the center of rotation of the binocular is the same as the center of rotation of your head.

4. The turning moment on a traditional tripod head increases as the elevation of the object that you are observing at increases.

Some solutions:

- Mirror mounts effectively solve or eliminate all of the issues above, but introduce new ones, including a reversed sky view, dewing, and difficulty in locating objects (unless you use a green laser or reflex finder).
- A *Starchair* solves all of the above, but introduces issues of storage, transport—and expense!
- Many DIY binochairs solve all of the above issues.
- A reclining observing position solves #1 above.
- Angled eyepieces solve issue #1 above.
- Tripods are useful for supporting anything you use to solve #2 above (e.g., a parallelogram mount or a lateral extension arm). Ideally this will be counterbalanced, or the tripod will be supporting a cantilevered load and there is a risk of tipping.
- An adjustable center post on the tripod solves #3.
- Parallelogram mounts solve #2 and #3 (and, if it is properly designed, #4).

My preferred solutions:

- Up to 10×50: handheld+recliner
- 15×70 straight-through: recliner+monopod+trigger-grip ball-head *OR* recliner+parallelogram
- 100 mm or larger: angled eyepieces+parallelogram

Bibliography

Chairs

Craig Simmons' Binocular Chair: http://www.cloudynights.com/photopost/showgallery.php?cat=500&ppuser=1895&password=
Starchair: http://www.starchair.com/

Mirror-Mounts

Florian's Binocular Viewing Accessories: http://www.stargazing.com/bino/index.html
Sky Window: http://www.tricomachine.com/skywindow/

Parallelograms

A Bino-Mount Built with Comfort in Mind: http://home.att.net/~jsstars/binomt/binomt.html
A Quick and Easy Binocular Stand: http://www.mdpub.com/scopeworks/binostand.html

Binocular Mount: http://www.astro-tom.com/projects/binocular_mount.htm
Building a Parallelogram Binocular Mount: http://home.wanadoo.nl/jhm.vangastel/Astronomy/binocs/binocs.htm
Parallelogram Binocular Mount: http://www.gcw.org.uk/bino/binonet.htm
Steve Lee's bino mount: http://www.aao.gov.au/local/www/sl/sl-tels.html
Tim Phizackerley's Binocular Mount: http://www.timphiz.co.uk/funstuff.htm
Universal Astronomics: http://www.universalastronomics.com/

Miscellaneous

CloudyNights Reviews of Binocular Mounts: http://www.cloudynights.com/mounts.htm

Chapter 7

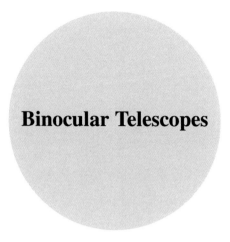

Binocular Telescopes

Binocular Telescopes

The distinction between "binoculars" and "binocular telescopes" is fuzzy, to say the least. Some people introduce a further complication by insisting that a third category, "binoscope," should be considered. Given that both "binocular" and "binoscope" are essentially contractions of "binocular telescope," it seems to me that any additional complication in terminology is more likely to be something that encourages dispute than something that adds any meaningful or useful clarification. That said, in this book, the term "binocular telescope" is usually, but not mutually exclusively, applied to binoculars that have one or more of the following characteristics:

- Larger than 150-mm (6″) aperture
- Focal ratio of f/5 or greater
- Use reflecting (usually Newtonian) optical systems
- Constructed from optical tube assemblies initially intended or sold as telescope tubes
- Use interchangeable eyepieces (especially those that are sold as telescope eyepieces)

The aim is usually to get either more aperture or a more forgiving focal ratio than is found in most binoculars (which typically have a focal ratio around f/4), thus allowing greater light gathering and the potential for higher magnification.

The majority are home constructed (Fig. 7.3), but there are also commercially available models (Figs. 7.1 and 7.2). Jim's Mobile Inc. has concentrated on developing Newtonian binoculars, while the Hutech Corporation has taken advantage of the modular design philosophy of Borg refractors to develop "binoscope" solutions.

Fig. 7.1 250-mm (10″) aperture binocular telescope by JMI. The picture shows the inside of the optical tubes (Photo courtesy of Jim's Mobile Inc.)

Fig. 7.2 Borg 125SD (Haruka). Borg binocular telescopes have parts that are interchangeable with each other and with conventional Borg telescopes (Photo courtesy of Hutech Corporation)

Fig. 7.3 Binocular telescope by Peter Drew constructed from two 150-mm (6″) *Synta* telescopes

Some of these solutions are now transferable to other refractors such as Vixen and Takahashi.

The vast majority of binocular telescopes are made with either refractors or Newtonian reflectors. Catadioptric telescopes such as the Schmidt-Cassegrain (SCT) and Gregory-Maksutov-Cassegrain (Mak, MCT) telescopes are unsuitable due to the focusing method used in the majority of the commercially available ones. They focus by moving the primary mirror and all suffer from a degree of mirror "flop" as a result of this. In manifests itself in two ways. Firstly, the mirrors shift their orientation slightly when the instrument is being focused. Secondly, they can shift when the telescope changes its orientation. When they are used as equatorially mounted telescopes, this typically happens when they move from one side of the meridian to the other. Although binoculars are rarely mounted equatorially, mirror flop could still occur when the instrument is moved. Of course, no two mirrors will

Fig. 7.4 Binoscope *Binobacks*. *Binobacks* offer a simple and flexible solution for connecting eyepieces to a pair of telescopes (Photo courtesy of Binoscope)

Fig. 7.5 The focusing, IPD, and collimation mechanism in the *Synta*-based binocular telescope. Collimation is achieved by adjusting the elliptical mirrors

shift in exactly the same way, so collimation becomes a nightmare! The suggested solution is to fix the primary mirror and attach a conventional focuser, such as a Crayford or a helical focuser, to the visual back of the telescopes. There are no commercially made binocular telescopes of this type, and very few amateurs have attempted it.

When these telescopes are home constructed, it is crucial to have a good focusing mechanism that also allows for collimation (Fig. 7.5). These instruments often work at higher magnifications than equivalent binoculars and thus collimation tolerances are significantly more severe. Typically, they have to be recollimated every

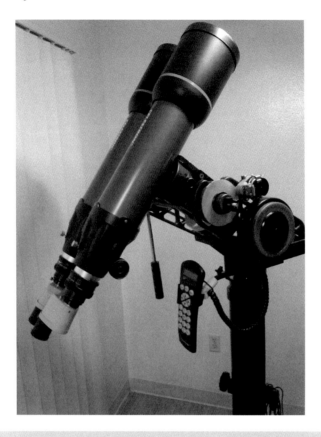

Fig. 7.6 **Dual 102-mm f/6 Celestron Binoscope** (Photo courtesy of Norman Butler)

time they are used, so ease of collimation is a must. A significant advance in this regard was the development of the Erecting Mirror System by the Japanese optician, Tatsuro Matsumoto. This is the heart of his *Binoback* (Fig. 7.4). A pair of *Binobacks* can merely be inserted into the focusers of a pair of parallel-mounted telescopes and adjusted to enable the eyepieces to be placed at the correct interpupillary distance for the observer. There are adjusting screws on the mirrors to enable precise collimation of the system. It is, of course, essential that the telescopes are held parallel and a number of manufacturers make saddle plates for this purpose.

Norman Butler uses a simpler method in his Dual 102-mm f/6 Celestron Binoscope (Fig. 7.6). This binoscope, which won the Warren Estes Memorial Merit Award at the 2011 Riverside Telescope Makers Conference, is mounted on a GOTO mount with a 40,000 object database.

Collimation is achieved with simple X-Y adjustments on the refractor mounting base platform. Butler reported that the biggest challenge he faced was how to adjust for the interpupillary distance without introducing miscollimation. His very simple, but also very effective, solution is a pair of microscope eyepiece holders. IPD is changed by swinging one or other (or both) of the eyepiece holders.

Fig. 7.7 **Bruce Sayre's 22-in. binocular** (Photo courtesy of Bruce Sayre)

For those wanting large-aperture binocular telescopes, the expense of purchasing two high-quality optical tube assemblies and mounts that can both handle the weight and be fitted with an appropriate saddle plate has been an impetus for amateur telescope makers to take up the challenge.[1]

One of the first successful very-large-aperture binocular telescopes is the 22-in. f/5 binocular that was completed by Bruce Sayre in 2003 and which won a Merit Award at the 2004 Riverside Telescope Makers Conference (Fig. 7.7).

According to Sayre,[2] the key features of his superb instrument include:

- *22" f5 Newtonian binocular telescope on an alt-az mount.*
- *9½' tall, with center of gravity 16" above ground to minimize eyepiece height.*
- *Economical use of materials through a minimalist design.*
- *338-lb total weight, including lead counterweights.*
- *48-lb, 1 5/8" thick primaries, 4 1/2" secondaries, 2" tertiaries.*
- *Mirrors are glued to triangular cells to simplify cell structure.*
- *Strut tubes are only 5/8" in diameter and use midpoint bracing to prevent sag.*
- *Spiders are made with 12 strands of .02" wire.*

---

[1] See, for example, http://www.binoscope.co.nz/links.htm
[2] http://www.brucesayre.net/#Overview

- *C-rings (or trunnions) are mounted inboard to minimize azimuth and ground ring diameter.*
- *Adjustments to converge both sides into optical parallelism are separate from— and do not compromise—collimation.*
- *2" eyepieces are supported with a wide range of interpupillary adjustment.*

He has also fitted a drive system based on the *Sidereal Technology*® servo motor controller, in which brushed DC servo motors drive the telescope via friction rollers. Altitude and azimuth are sensed by optical encoders. The drive/tracking system is powered by a 12-V gel cell battery and can interface with a wide variety of planetarium programs and apps as well as dedicated astronomical computers like the *Argo Navis*®.

He typically uses 30-mm eyepieces, which give a magnification of ×93 and an apparent field of view of 80°—the advantage of a large binocular over a single telescope with an equivalent aperture (in this case, 31″) becomes apparent with this sort of combination. It is here that the *Binocular Advantage*, outlined in Chap. 1, really shows itself. The benefits of increased visual acuity and contrast sensitivity, combined with a lower photon-detection threshold, enabled the observer to see more color, detail, and structure in deep-sky objects. For the visual observer, the experience of a large-aperture binocular is second to none!

Recently, amateurs have made significant innovation in the construction of binocular telescopes. This is exemplified by Keith Harlow's 16″ Newtonian binocular, which includes the following characteristics:

- Primary mirrors seated on a removable (for transport) box-section frame.
- Focusing by moving the primary mirrors (Fig. 7.8) and includes a backlit display of focus position, which simplifies the repetition of a focus setting for different eyepieces or different users.
- Servo motors give lateral movement of eyepieces to permit different interpupillary distances and diopter adjustments to be made. This has three storable settings.
- Removable (for transport) eyepiece section with tapered dowel pins for accurate relocation.
- Fine adjustment of all primary, secondary, and tertiary mirrors (Fig. 7.9), essential for fine collimation.
- "Swingable" secondary mirror to direct the optical axis to a camera (Fig. 7.10).
- *Lumicon NGC Sky Vector* ® "pushto" digital setting circles.
- Electrically height-adjustable pier to compensate for different eyepiece heights, either for observing at different altitudes or for different observers.

Harlow reports that it took several days of work with lasers and bubble levels to get the optical axes of each side aligned both with each other and with the focusing axis. Without this, collimation would have been all but impossible. He has also found, through trial and improvement, that the most comfortable eyepiece setting has them "coned" in about 2° each; he found it tiring when they were perfectly parallel. Although there are inevitable problems with thermal currents from the

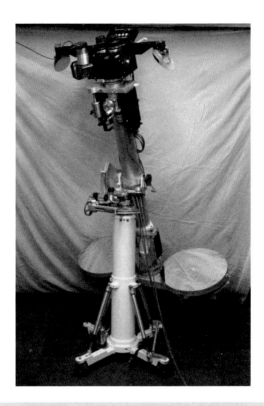

Fig. 7.8 Keith Harlow's 16″ Newtonian binocular (Photo courtesy of Keith Harlow)

Fig. 7.9 Fine adjustment mechanism of a secondary mirror (Photo courtesy of Keith Harlow)

Binocular Telescopes 127

Fig. 7.10 A secondary mirror swings to enable the use of a camera (Photo courtesy of Keith Harlow)

observer degrading high-magnification images of the planets, the solution may be a deflector system, but Harlow says this is not a priority. Overall he reports, "It's quite hard to describe just how much better it is than squinting down a 'mono' scope. 'Spacewalk' is a word often used by binocular telescope users. I think that probably sums it up nicely".

Chapter 8

Observing Accessories

Finders

You can easily aim straight-through binoculars without a finder as long as the magnification is below about ×20, when there is a real field of view of 2.5° or more. Aiming becomes more difficult at higher magnifications and can be extremely difficult with angled eyepieces unless you are exceptionally familiar with the *binocular* view of the region of sky you are observing. Some form of finder is therefore extremely helpful in these latter cases. In order to avoid fogged optics, you should ensure that you mount the finder so that, when you use it, you are not breathing on the eyepieces of the binoculars. There are four distinct options:

Simple Mechanical Sight. A simple mechanical sight usually takes the form of either a sighting tube or a "vee and blade" sight. Both of these are amenable to DIY construction. They can be mounted to the binocular with adhesive hook-and-loop (e.g., *Velcro®*) strip (loop on the binocular, hook on the sight) or rubber bands.

"Vee and blade" sights were common on large naval binoculars up until the 1950s. Of all the options for binocular finders, these are probably the simplest sight to use. You can make a simple "vee and blade" sight with a strip of metal approximately 25 cm × 1 cm × 3 mm (10 × 1/2 × 1/10 in.). One end has a notch filed into it and the other is either filed to a point or, preferably, twisted 90° about a longitudinal axis. About 5 cm (2 in.) of each end is bent at right angles in order to form a "U" shape. You mount the sight on the binoculars with the vee nearer the eyepiece end. You can align it with the optical axis of the binoculars merely by bending the metal strip. If you use a "vee and blade" you should ensure that there is no possibility of it coming into contact with your eye!

A sighting tube is merely a straight tube of metal or plastic, about 15–20 cm (6–8 in.) long with a bore of about 1 cm (½ in.) diameter. Such a tube will have a field of 2 or 3°, depending on how far in front of your eye it is. It is therefore relatively trivial to position it so that its field is approximately the same as that of the binoculars. It can be less trivial to align it with the optical is of the binocular, but this is usually possible with a little ingenuity aided by trial and improvement. The near end should not protrude beyond the eyepiece and, for added safety, should be edged with some soft material such as foam tape or have a something like a soft rubber tube pulled over it to form an eyecup. Compared to a "vee and blade" sight, the tube is simpler to store as it can usually be slipped into the binocular case.

Reflex Finders. Reflex finders are unit power (i.e., no magnification) devices that use a simple optical system to project the illuminated image of either a dot or a reticle of concentric circles onto the sky. The red dot finders tend to be more compact than the reticle type and you may find that you can store it in the binocular case. Reticle finders are more bulky but are usually considered to be more useful, especially if the diameter of one of the circles is a close match to the field of view of your binoculars. Whether or not the circles match the binocular field, you can use the circles for precise star-hopping.

Reflex finders usually include some form of aiming adjustment. In the "red dot" finders, this is usually achieved by knurled-head screws that move the finder relative to its base. In reticle finders, it is usually the orientation of the reflecting surface that is varied to change the aim. Most include a dimmer switch to alter the brightness of the dot or reticle so that it can both be seen against a bright sky and not "drown out" faint objects in a dark sky. The base of the finder clips into a mounting bracket that is attached to the binocular. Once I have established the optimum position for the finder, I fix the mounting bracket to the binocular with double-sided adhesive foam pads. Some manufacturers provide these pads with the finder but, if this is not the case, you may obtain either pads or double-sided adhesive foam tape from a good stationery or hardware store. These finders may be provided with two mounting brackets so that they can easily be swapped between instruments; alternatively, spare mounting brackets can usually be obtained via the supplier. To avoid breathing on the eyepieces and fogging them when you use the finder, if you use your right hand eye with a finder, you should mount the finder on the left tube and vice versa.

It is a common misconception that reflex finders are used only with one eye looking through it. Although this mode of use is possible with bright objects, if you are looking through the reflecting surface all objects appear dimmer and most stars disappear entirely. The correct mode of use, as with all straight-through finders, is to begin with both eyes open. The eye that is not looking through the finder gets an unattenuated view of the sky, and your brain merges the images received by both eyes, exactly as it does in "normal" use of a pair of Mark I eyeballs!

My preferred reflex finder is the *Rigel Quikfinder®*, which is relatively compact and whose aperture stands some 75 mm (3 in.) from the body of the binocular (Fig. 8.1). This sight also has the advantage of the facility to make the illumination blink on and off at an adjustable rate; this can be a great aid when targeting an object that is sufficiently faint to have its light obliterated by the finder illumination.

Fig. 8.1 *Rigel Quikfinder®* reflex sight

Most reflex sites are very prone to dew. Proprietary dew shields are available, but it is a simple matter to make one with sticky-back hook-and-loop tape and 2-mm foam sheet or similar material.[1] Since I made the one shown in Fig. 8.4, I have had no dew problems with the finder, even on nights where the entire binocular and dew shield were dripping wet with dew.

Finder Scopes. Some people prefer finder scopes to unit power finders and some large binoculars come equipped with them. The scope need not be of high power; that provided with my 100-mm *Miyauchi®* is a 3×12. If the finder is to be mounted between the optical tubes of the binocular, it should have a significant amount of eye relief if you are going to be able to avoid breathing on the eyepieces when you use the finder. The eye relief of the finder mentioned above is 65 mm. This amount of eye relief is not to be found in finders intended for use with telescopes and these must consequently be mounted to one side in the same manner as suggested for reflex finders (above). On the other hand, telescopic sights designed for rifles do have adequate eye relief and have a magnification more in keeping with what is required for binoculars, although the field of view may be somewhat small. Although telescopic finders are useful in the daylight, I find them to be inferior to reflex sights for nighttime use with binoculars, but this is obviously a matter of personal preference (Fig. 8.2).

Lasers. Over recent years the price of green laser pointers has been decreasing and there is a growing trend of using them for astronomy, both as pointers and as finders. With the single exception of mirror mounts, for which they are the most practical type of finder, any advantage they offer over reflex finders is, in my opinion,

---

[1]http://astunit.com/atm.php?topic=quikfinder

Fig. 8.2 3 × 12 finder scope with 55-mm eye relief

more than offset by the combination of cost, potential danger of eye damage, general nuisance to other astronomers, and potential danger to aircraft. If you do use a green laser in company, you should exercise extreme care and should ascertain that none of the company objects to its use. You should also find out what statutory regulations exist in your country, regarding civilian use of lasers outdoors.

Filters

The most useful filters are a UHC or an O-III filter. If you are only going to get one, get the UHC. The standard 31.7 mm (1.25 in.) filters sold for telescope eyepieces can be used with most large binocular eyepieces, merely by being held over the eyepiece. It is not ideal in this position, as it is designed for placing near the focal plane, but it is certainly effective. If you have a binoviewer or a binocular telescope, then you obviously use them as intended. There is no need to acquire two—a good method of using it is to "blink" between each eyepiece, when the sought after nebula often becomes obvious. An O-III filter is especially useful for identifying those planetary nebulae that appear to be stellar at the magnification of the binocular.

Also useful, especially for solar eclipses and transits, are solar filters. These can be simply made, to fit over the objective lenses, from *Baader AstroSolar*® film. Solar-filtered binoculars are particularly useful for group viewing of solar phenomena (Fig. 8.3).

Fig. 8.3 Group solar observing with filtered binoculars

Dew Prevention and Removal

In order to know how to combat dew, it is important to have some understanding of why it forms. Water vapor condenses out of the air onto any surface and simultaneously evaporates from that surface. The potential rate of evaporation is lower at lower temperatures. Below a specific temperature, the dew point, the rate of evaporation is lower than the rate of condensation and dew forms. The principles of dew reduction are then simple: reduce the amount of cooling of the optical surfaces and reduce the amount of warm moist air (especially breath!) that comes into contact with them.

Under a clear sky, objects, including optical surfaces, lose heat by radiative cooling. Outside our biosphere is space; far enough out and it is space at a temperature of 2.7 K. Although the effective temperature of the sky is perhaps 100 K or so warmer than that, it is still a great deal colder than the surface of the Earth. Hence, on clear nights (i.e., those good for astronomy), there will be a net loss of heat by radiation from the surface of the earth and things on it, like binoculars. As they cool, they become prone to dew (and frost) formation.

Our simplest way of reducing dew formation is to reduce the amount of sky which the optical components can "see." Binocular objectives and reflex finders are among the most dew-prone of all astronomical surfaces. Dew shields provide the simplest way of shielding binocular objectives from the cold sky but, of those binoculars that do have slide-out dew shields, very few are provided with ones that are sufficiently long. To be fully effective, a dew shield should be at least 2½ times

Fig. 8.4 Dew shields. Simple "passive" dew shields can be made for binoculars and reflex finders using foam sheet

as long as the aperture it is shielding; four times is preferable and almost always effective. An extension of this length is unwieldy on small and medium binoculars, for which there are simpler methods for dew prevention and removal. Simple dew shields for larger binoculars can be made from stiff plastic (that from plastic "wallet" folders is usually adequate) or 2-mm-thick foam sheet. These can be stored flat and can have their edges secured in use with self-adhesive hook-and-eye (*Velcro®*) strip. The binocular in Fig. 8.4 has 100-mm objectives. The slide-out dew shield has a length of 6.5 cm; the foam extension extends 25 cm beyond the front of the lens and is far more effective at dew prevention. Since I made it, I have not had an observing session cut short by dew on the objectives.

For those who want a higher-tech solution, there are proprietary dew heaters, such as the *Kendrick Dew Zapper®*, that are available commercially. These provide a low-level heat to the surrounding of the aperture. A DIY alternative, if you have the requisite skills, is to make a similar device using resistance wire or strings of resistors taped to the surround of the aperture. These need not impinge on the light path. Those readers with electronic capabilities will, no doubt, be able to see more sophisticated solutions.

For small and medium binoculars, the solution I use nowadays is to hang the binocular inside my jacket as soon as there is any sign of dewing and, on cold nights, when I am not actually looking through them. If you do this, you will find that they immediately dew up even worse from the warm moist air under the jacket, but they soon clear and are ready for use again. Because handheld binoculars are usually not held to the eyes for very long periods, their objectives tend to cool less quickly and they are not as prone to dewing as are mounted binoculars.

Eyepieces on larger binoculars offer a different problem. For obvious reasons, a long dew cap is not an option (and eye cups even make the matter worse!). The obvious thing is to avoid breathing on them, but there is another source of warm moist air: our eyes. It makes sense to dry a moist eye before putting it to an eyepiece, particularly if that eyepiece has an eyecup, which will trap any moist air.

Fig. 8.5 12-V hair drier and battery pack

On particularly cold nights, fold down or retract the eyecup. There are two obvious ways of warming eyepieces: an inside pocket or some form of electrical heating. I have never tried the latter (but there are commercially available eyepiece heaters), but I routinely swap eyepieces when I am observing in winter with my 100-mm binoculars.

The practical alternative to dew prevention is dew removal. Several astronomical suppliers provide "dew guns" that are merely 12-V portable hair driers that plug into the cigarette-lighter socket of a car or battery pack. Exactly the same item is usually significantly less expensive if it is obtained from a camping store as a "travelling" hair dryer (Fig. 8.5).

Compass

A compass is invaluable when you are seeking twilight objects, be it Mercury at elongation or the evening objects during a Messier Marathon. A simple hiker's compass with a bezel that can be adjusted to compensate for local magnetic declination is ideal. If it has a tritium-lit luminous needle, north point, and plate arrow, so much the better.

Charts and Charting Software

Our need, as binocular users, for sky charts, is no less than the need of telescopic astronomers but is slightly different. Unless we are using giant binocular telescopes, we do not need charts that go as deep as those preferred by users of large telescopes. This means that our needs are usually completely met by the better "paper" charts and by most of the commonly available star-charting software. For example, the excellent *Sky Atlas 2000* goes down to magnitude 8.5 and incorporates galaxies and nebulae that are fainter than this. The choice of these is therefore a matter of preference and, often, familiarity.

If our choice to use binoculars is based to some extent on their extreme portability, we may wish to use charts and/or software that incorporates the same philosophy of choice. If this is the case, there is one paper chart that stands out: *Collins Gem Stars* (some older editions were called *Collins Gem Night Sky*). This little book is small enough to fit into a shirt pocket and contains sufficient information to keep the users of small and medium binoculars amused for many nights.

If you prefer to use charting software, the "extremely portable" route suggests using a handheld computer/personal digital assistant (PDA) or smartphone or, if you prefer something larger, a tablet computer. There are a number of excellent software options for these, depending on the operating system used by the handheld device. In general, for astronomical software, Palm OS is better catered for than other PDAs. There is a variety, and increasing quantity, of excellent astronomical apps for smartphones and tablet computers available on both *Android* and *iOS,* the two most common operating systems. The ones I find most useful include the following.

Palm OS. Of the many examples of astronomical software available for the Palm, there are three planetarium programs that stand out:

2Sky: This is commercial software and is not offered as an evaluation version but has a 30-day refund policy if you find the software to be unsuitable. The "basic" version ($25) has stars to magnitude 7 and 500 deep-sky objects (DSOs), "total" version includes stars down to magnitude 9.5 and 13, 600 DSOs from the Messier/NGC/IC catalogues, and the "mega" version has stars to magnitude 11.2 and the same DSOs as the "total" version. It also comes with *2Red*, which changes the entire PDA to red-screen night mode.

Planetarium: This is "nagware," i.e., shareware which, until you register it, reminds you when you start and/or close the program that it is unregistered. It costs $24 to register. This is my most-used astronomical software. It has a "Compass View" that shows the lunar phase and, at a quick glance, the altitude and azimuth of the Sun, Moon, major planets, and one other object of your choice. It has instantly accessible rise and set tables and twilight tables. Among its most useful features is the ease with which catalogues of your choosing can be added to its database and with which objects can be imported into a "Personal" catalogue that can be exchanged with other *Planetarium* users. It has stars to magnitude 6.5 as standard, but there are databases that go, by increments of one magnitude, down to 11.5 (i.e., the entire Tycho2/Hipparchos catalogue).

PleiadAtlas: Like *Planetarium*, this is nagware ($10 to register). It goes down to magnitude 11.5 and incorporates the Messier, NGC, and IC catalogues.

Android and iOS

SkySafari (Android and iOS)*:* This is, at the time of writing, by far the most capable piece of astronomical software for Android, having most of the functionality of good desktop software. Additionally, it can use the device's compass and gyroscope to help you identify objects in the sky. It comes in three versions, two of which are suitable for binocular observers:

- *SkySafari* ($2.99) has 46,000 stars, plus 220 deep-sky objects. Good entry-level software.
- *SkySafari Plus* ($14.99) has 2,500,000 stars and 31,000 DSOs, probably more than sufficient for any binocular observer, including those with very large binocular telescopes. For these fortunate observers, it will also control several kinds of astronomical mounting.

SkEye (Android)*:* A planetarium app that has a particularly good implementation of "PUSHTO" functionality. Whereas most other apps that use the compass/gyroscope will show what sky is behind the device, SkEye allows you to mount the device at any angle you wish on your observing instrument, then configure the app so that it shows what the instrument, not the back of the device, is pointing at. It also warns you if your environment or equipment is producing strange magnetic fields that may interfere with the compass. The basic version (free) includes the Messier catalogue and approximately 180 NGC objects; the Pro version (£5.53) has the entire NGC and IC.

LunaSolCal (Android and iOS)*:* A very comprehensive (free on Android, $1.99 on iOS) lunar and solar calendar app that enables you to plan observing sessions with its comprehensive output of sun- and moonrise and sun- and moonset, lunar phase, altitude and azimuth, twilight, and a host of other features. It is loses accuracy at latitudes higher than 65°.

Astro Panel (Android)*:* Provides you with a 3-day astronomical weather forecast based on your device's GPS location, including cloud cover (in okta), temperature, transparency, seeing, humidity, and state of the Moon.

Sky Harbinger (iOS)*:* Similar to Astro Panel for Android.

There is one caveat to using handheld device: dark adaptation. Even if you do use a "red mode," there is usually still sufficient light to destroy your eyes' dark adaptation. The solution is to cover the screen with red translucent plastic sheet. If your device has a capacitative touch screen, you should make sure that it will still work through the plastic sheet. Remove it if you have an emergency need for a torch (flashlight).

Torches (Flashlights)

First, a note on terminology, born of several misunderstandings I have encountered when communicating across the Atlantic. In UK English a "torch" does not have the same meaning as in American English. It does not mean a "flaming torch," but

Fig. 8.6 Small accessories. *Top L-R*: Compass, pocket star atlas, illuminated magnifier. *Middle*: Handheld computer running Planetarium® software, red LED torch (*flashlight*) in *Nite Ize®* headband. *Bottom*: White-light torch (also fits in headband) for use when setting up and dismantling kit

what is, in American English, called a "flashlight." A red-light torch is a useful piece of observing kit, not only for reading charts but also for examining equipment when this is necessary. Torches that use red-light emitting diodes (LEDs) are more than adequate for most purposes. I find that they are much more useful if they can be head-mounted, thus leaving both hands free. LED head torches are now commonly available but, for those who prefer the robustness of a good-quality small handheld torch such as a *Mini Maglite®*, they can be head-mounted with an adjustable headband called a *Nite Ize®* that is made for precisely this purpose.

There is, however, a potential problem with red illumination for reading charts. Red light focuses further behind the retina than yellow light and, especially those of us who eyes have become presbyopic with age, it can be difficult to focus the red-illuminated page. A solution, which is also applicable to people who prefer to observe without spectacles but who must use them for reading charts, is to use an illuminated magnifier. The light source can be covered with translucent red plastic or with red nail varnish. The magnifier is also useful on a PDA screen (Fig. 8.6).

A white-light torch is useful when assembling and dismantling observing apparatus.

Storage and Transport Container

It makes sense to keep, with your binoculars, the various small items that you frequently use when observing. There are numerous options for containers and some binoculars come with very useful cases, although I have yet to find one that

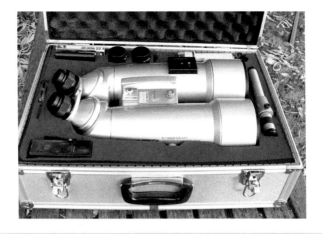

Fig. 8.7 Binocular storage case

is ideal. A relatively inexpensive option is an aluminum camera case. If it comes with soft polyurethane foam, this should be replaced with a high-density polyethylene foam. This can easily be cut and shaped with a carving knife or a sharp chisel. Recesses in it can be made for torches, charts, finders, spare eyepieces, etc. (Fig. 8.7).

Cases supplied with medium and small binoculars have very little capacity for extra storage. However, in my 10×50 case I do also keep the L-bracket for tripod mounting the solar filters (which I store on the objectives) and a small planisphere.

Software Sources

In addition to the relevant App Store and Google Play sources that come preinstalled on modern devices, there is more information on the software/apps mentioned in this chapter at:

2Sky: http://open2sky.sourceforge.net/
Planetarium: http://www.aho.ch/pilotplanets/
PleiadAtlas: http://www.astronomycorner.net/PleiadAtlas/
SkySafari: http://www.southernstars.com/
LunaSolCal: http://www.vvse.com/products/en/lunasolcal.html
SkEye: http://lavadip.com/skeye/
Astro Panel: http://astrotips.com/software/astro-panel
Sky Harbinger: http://www.sibimon.net/node/7

Chapter 9

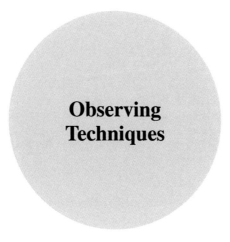

Observing Techniques

As any observer with any experience will know, using appropriate techniques will not only increase the chances of observing difficult objects but will also enhance the pleasure of observing.

Personal Comfort

All other things being equal, the efficacy of visual observation is usually in direct proportion to the comfort of the observer. For the binocular observer, this means three things: physical comfort, thermal comfort, and nutritional comfort.

Physical Comfort. The plethora of homemade observing aids ranging from simple recliners to computerized observing chairs is an indication of the degree of comfort that binocular observers seek to find. This should come as no surprise: the simple fact that one has chosen binoculars is often an indication that one has made a choice of optical comfort!

For those of us who prefer to stand while observing (usually restricted to those of us whose binoculars are mounted and have angled eyepieces), our comfort is relatively easy to attain. On the other hand, those of us who prefer some sort of seating or reclining may well find it considerably more difficult, owing to the differences in our backs and necks. That which perfectly suits one person may well be an anathema to another. My best advice is to try out as many options as you can and, once you have found something that suits you, acquire or make it and then treasure it. My *Mac Sports*® recliner is such an item, and I treat it with almost as much care as I do my binoculars themselves. If you find something close to what is ideal for you, try to adapt it. This is usually achievable by judicious use of cushioning or padding, so that your back and neck are perfectly cradled, and your arms are supported but not restricted.

Thermal Comfort. Almost by definition, astronomical observing takes place during the coldest part of the day. Each of us not only has a different sensitivity to and tolerance of cold, but this individual variation itself varies from day to day and even during the course of a night's observing. Therefore, our personal insulation needs to be both adequate and adjustable; this implies layering. Furthermore, attending the eyepieces of the binocular is not a physically active task and, for this reason alone, we need to dress as though it was about 5–10°C colder than it really is in order compensate for the lack of physical activity. Those who are used to being in cold conditions will undoubtedly have their own clothing solutions. Those who are not may find some useful advice in the following advice, whose principles have kept me comfortable for over a decade.

- *Undergarments*: If you want to keep warm, eschew cotton; it is cold because it absorbs moisture (which is why we use it for towels) and uses our body heat to evaporate the water. If you wish to stick to natural fibers, change to wool (itchy) or silk (expensive), otherwise you need to use a synthetic "thermal" hydrophobic fabric that wicks water away from the skin without absorbing it.
- *Insulting layer*: The middle layer of clothing is the insulating layer which needs to trap as much air as it can, because air is an excellent insulator. By far the most efficient insulating layer per unit weight, or per unit volume, is dry goose down, but the damper it gets the less effective it is. It takes ages to dry out, and it is extremely expensive. The moisture that has been wicked through the inner layer comes to the middle layer, which it will dampen unless it passes through. If you wish to stick to natural fibers, wool has the reputation for being a good insulator when it is wet, although it can be a bit heavy. Modern synthetics such as *Hollofil®*, *Thinsulate®*, and *Polartec®* are excellent insulators and will wick moisture away from the body without absorbing it. It makes sense to have a zipped garment as this middle layer, so you can adjust your insulation to different conditions by opening and closing the zip. Also remember that this layer should not be too tight or be compressed by the outer layer, or its insulating properties (related to the amount of trapped air) are reduced.
- *Outer layer*: While we tend not to observe in strong winds, even an 8-km/h (5 mph) breeze can make a great deal of difference to our thermal comfort if it can get into the insulating layer. The outer layer therefore needs to be windproof, but it must also pass the water vapor which has been wicked away from our bodies by the inner layers. A windproof *Polartec®* fleece is ideal as a combination insulator/outer.
- *Hats*: It is said that we can lose anything up to 40 % of our body's heat through our heads. Even if this is as low as 25 %, it indicates that we can regulate our body temperature by changing our headware, thereby reducing the need to fiddle about with the insulating middle layer of clothing. "Extreme conditions" headware would follow the same pattern as our other clothing, i.e., silk or polypropylene balaclava, covered by a wool or *Polartec* layer, covered by a windproof

layer but, for most of us, a simple fleecy hat, preferably with earflaps, is adequate.
- *Gloves*: There is no particularly elegant solution to the need to keep the fingers warm and also have them free and sufficiently sensitive to make fine adjustments. The best solution I have found is an insulated "hunter's" glove that has a fold-back mitten over a fingerless glove as well as split thumbs. This keeps the fingers toasty warm, but enables the forefinger and thumb to easily slip out when necessary.
- *Footwear*: On clear nights the ground cools faster than the air and we lose heat rapidly to the cold ground if we wear thin soles. Ordinary thick-soled shoes worn with two layers of socks (inner "wicking" sock, outer insulating "cushion loop" sock) are sufficiently warm for most temperate zone conditions, and for the more cold-footed among us, there are alternatives such as snow boots, which have thick soles and a removable inner sock of thick felt, often combined with a sandwiched reflective layer.

Nutritional Comfort. Adequate nutrition is essential as it combats tiredness and cold. It is difficult to observe comfortably after a heavy meal and is equally difficult when we are conscious of hunger. If we are cold, we need to replace our lost energy with carbohydrates. It is usually healthier to take these in the form of starches, which release their energy slowly, than as a sugar, although those who do not have contraindicating dietary requirements may find a warm glucose-based drink to be beneficial in some circumstances. It is also essential to keep a good fluid balance. Time can pass very rapidly when we are enjoying observing, and our steamy breath on cold nights is an indication of fluid loss. Over the last decade or so, people have become more conscious of the need for adequate hydration but there is an additional benefit that is less well known: if we maintain good hydration, even by merely sipping cold water, our extremities tend to get less cold.

Observing Sites

Where possible, choose your observing site with care. Stray light is obviously to be avoided (Fig. 9.1). Also avoid observing over buildings or other sources of heat. Altitude can be an advantage as it can take you above sources of stray light and you have less (polluted) atmosphere between you and your target objects. As little as 300 m (1,000 ft) can make a noticeable difference; transparency is usually considerably better from, say, the North Downs of Kent than it is a few miles away on Romney Marsh. Conversely, if you cannot get to high ground, you may be able to observe over a sea horizon. The choice sites are those of very high altitude with a sea horizon, but few of us have access to such places.

If you observe on your own property, you can of course prepare it. Screens can be used to block intrusive lights if they cannot be occulted by buildings or vegetation. Equally important is to prepare the ground. You cannot observe in a relaxed manner if you are in danger of tripping over objects on the ground!

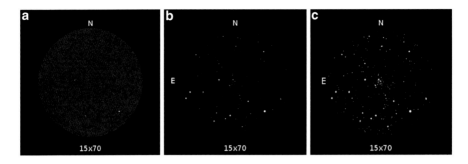

Fig. 9.1 A dark site can make a tremendous difference to what is visible. This is the Open Cluster M35 (Gemini) with 15×70 under (**a**) bright urban, (**b**) good suburban, and (**c**) dark rural skies

If you use a site away from home, do reconnoiter it in daylight and, where appropriate, get the landowner's permission. Take personal security into account. You cannot enjoy observing if you are concerned about the potential presence of domesticated or wild animals or of antisocial people.

Observing Techniques

Once our personal comfort is attended to, we can get down to the business of actual observation. There are various ways in which this can be enhanced:

Dark Adaptation. It takes at least twenty minutes to half an hour for our eyes to get properly dark-adapted, and the process of dark adaptation continues for over an hour. Some observers recommend closing your eyes for a few moments immediately before attempting a particularly difficult observation.

Averted Vision. Look down towards the tip of your nose so that light from the eyepiece appears to come from the "upper outside" of the eye. Use averted vision even for things that you can see well without it as this will very often allow more detail to be seen.

Perfect Focus. Ensure that your binoculars are perfectly focused. This will allow dimmer objects and more detail to be seen. Center-focus binoculars tend to get defocused more easily than those with independent eyepiece focus.

Keep the Binoculars Still. If something is difficult to observe, ensure that the binoculars are firmly mounted. Keep looking and sometimes the object just appears.

Jiggle the Binoculars. This seems contradictory to the advice above, but it sometimes helps, when using mounted binoculars, to examine the appropriate region with averted vision and to tap the binocular. The slight movement sometimes makes a difficult object appear.

Breathe. It is a normal reaction for us to hold our breath when we are doing something critical, especially if that activity is assisted by stillness. Try to overcome this tendency if you have it: a well-oxygenated retina is more sensitive. In particular, carbon monoxide from smoke reduces the ability of the blood to carry oxygen.

Patience and Persistence. These are probably the most important attributes of the successful observer. It can sometimes take several minutes to make a fleeting, difficult observation, and it may take several attempts over several nights before the various conditions are just right to allow the observation to be made. A patient, persistent observer can see more than a less patient one with better eyesight!

Part II
Deep Sky Objects for Binoculars

Chapter 10

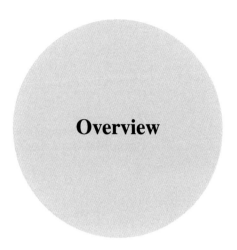

Overview

The objects in this part are ones that are visible with medium-sized binoculars, although some are visible in much smaller instruments. Most of the objects to the far south, for example, are easily visible in 8×30 binoculars. With the exception of objects that fill the eyepiece in 10×50 binoculars, all of these objects are better, in the sense that more detail can be eked out, in larger instruments. Similarly, many objects in the list for 100-mm binoculars are visible—or at least detectable—in smaller instruments, but often no detail is visible. For example, under good conditions an experienced observer can see the ***Ring Nebula*, M57**, in 10×50 binoculars, but it is stellar in appearance.

A third of the objects here are open clusters. Any "best of" selection is bound to be somewhat subjective, and I acknowledge that I am particularly fond of them as objects for binocular and small telescope observation. I have tried to include representative objects of all classes, but the relative ease with which open clusters can be seen, especially in nonideal sky conditions, coupled with the really wide variety they display, accounts for their apparent overrepresentation.

In addition to stars, the following classes of object are included:

- **Emission Nebulae.** These consist of gas that is ionized by the energy radiated by nearby stars. The light is emitted as electrons are recaptured. Many appear red in photographs, but the brighter ones may have a greenish tinge when observed by eye. They are often the sites of star formation.
- **Planetary Nebulae.** So called by William Herschel because many of them appeared as discs (although only about 10 % appear circular), that is, like planets, in his telescopes. They are the debris ejected from a star as it passes through the red giant stage to become a white dwarf.
- **Reflection Nebulae.** They are made visible by reflected starlight and are thus always less bright than the star that illuminates them (unless the star is attenuated

by intervening dust). It is the tiny dust particles, not the gas of the nebula, that reflect the light. They appear blue in photographs.
- **Galaxies**: Huge "island universes" of hundreds of billions of stars held together by gravity.
- **Globular Clusters**: Very old dense balls of hundreds of thousands of stars, which form halos around galaxies. All globular clusters formed in the same way, so we can safely assume that the brightest ones in different galaxies are of similar brightness. This enables them to be used as "standard candles" for measuring distances up to about 9 megaparsecs.
- **Open Clusters**: Less densely packed groups of stars than open clusters; may contain from a few dozen to a few thousand stars which recently formed in the galactic disc. The stars in open clusters are typically young.

The Object Catalogues

- **C—Caldwell:** A catalogue of 109 objects, numbered (with the exception of the Hyades and NGC4244) in order of declination from north to south. It was devised in 1995 by, and is named for, the late *Sir Patrick Caldwell-Moore* (better known simply as *Patrick Moore*).
- **Cr—Collinder:** The Swedish astronomer, *Per Collinder*, studied the structure and distribution of open galactic clusters. His 1931 catalogue was an appendix to one of his papers.
- **IC—Index Catalogue:** This is actually a combination of two catalogues, published in 1896 and 1905, of nebulae and double stars. It was compiled by *Johan Dreyer*, a Danish astronomer who worked in Ireland (Parsonstown and Dublin). It is a supplement to his NGC.
- **M—Messier:** *Charles Messier's* catalogue (augmented by *Pierre Méchain*) of objects for comet-hunters to avoid as they sought the return of Halley's comet. It was first published in 1771 and was updated for 10 years.
- **Mel—Melotte:** *Philibert Jacques Melotte* was a British astronomer of Belgian descent; among many astronomical achievements, in 1915 he published his eponymous catalogue of open clusters.
- **NGC—New General Catalogue:** This was compiled by *Johan Dreyer* and published in 1888, based on observations by *William, Caroline and John Herschel*, and *James Dunlop*. It is a massive work (7,840 objects) but contains many errors, some of which still remain after several revisions.
- **St—Stock:** A catalogue of twenty-four open clusters, mostly in the environs of Cassiopeia, compiled by *Jürgen Stock*, the astronomer who chose the Cerro Tololo observatory site in Chile.
- **Σ—Struve:** This is a catalogue of double stars that was published in 1837. It was compiled by *Friedrich Georg Wilhelm von Struve*, a German-born astronomer who founded the Pulkovo Observatory near St. Petersburg.

Summary Charts

These summary charts show only those deep-sky objects for which there are descriptions and finder charts. You may use the right ascension of zenith table to determine which charts to use for a particular date and time. (The fifth day of the month was chosen because that is when an hour of RA is approximately at the zenith on the hour of local mean time.) The object lists that appear after the charts enable you to plan your observing by object type, binocular size, or constellation.

		Month											
		Jan (h)	Feb (h)	Mar (h)	Apr (h)	May (h)	Jun (h)	Jul (h)	Aug (h)	Sep (h)	Oct (h)	Nov (h)	Dec (h)
Local mean time	18:00	01	03	05	07	09	11	13	15	17	19	21	23
	20:00	03	05	07	09	11	13	15	17	19	21	23	01
	22:00	05	07	09	11	13	15	17	19	21	23	01	03
	00:00	07	09	11	13	15	17	19	21	23	01	03	05
	02:00	09	11	13	15	17	19	21	23	01	03	05	07
	04:00	11	13	15	17	19	21	23	01	03	05	07	09
	06:00	13	15	17	19	21	23	01	03	05	07	09	11

Approximate right ascension of zenith on fifth day of month (add 1 h to RA every 15 days)

North Polar Region

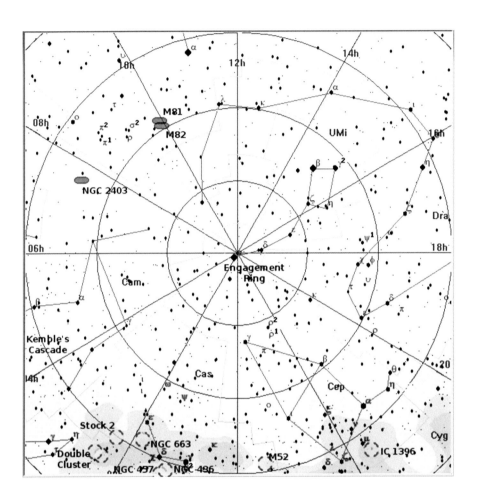

North RA 22 h 30 m to 01 h 30 m

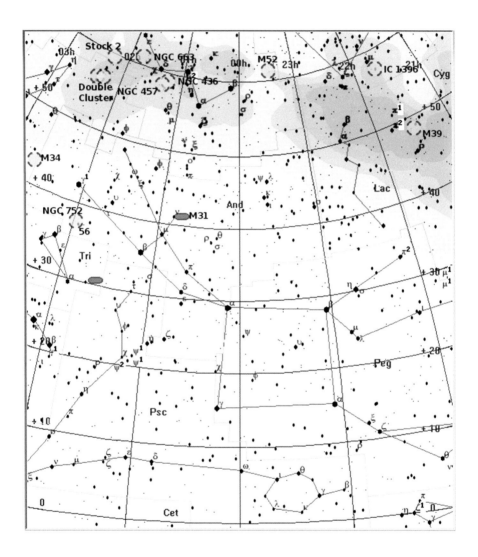

South RA 22 h 30 m to 01 h 30 m

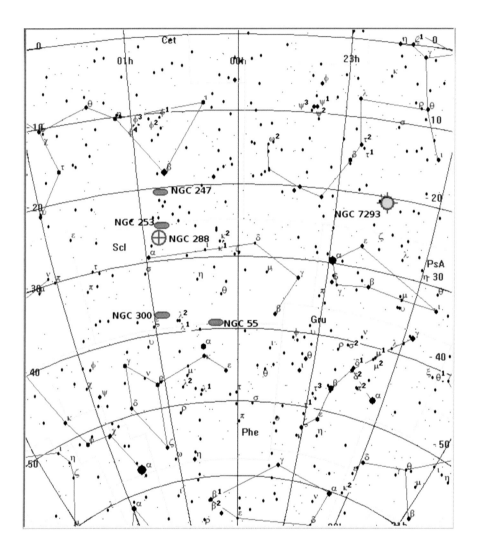

North RA 01 h 30 m to 04 h 30 m

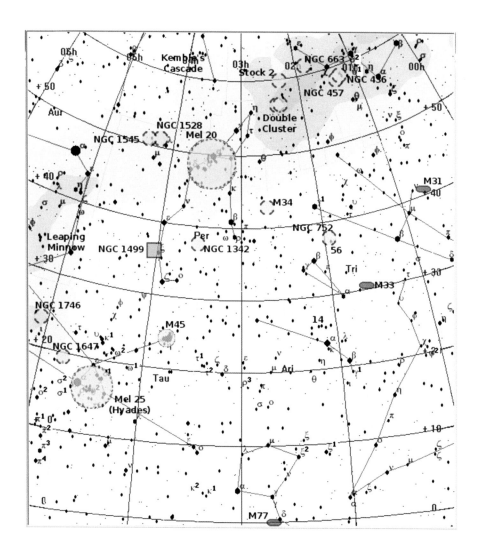

South RA 01 h 30 m to 04 h 30 m

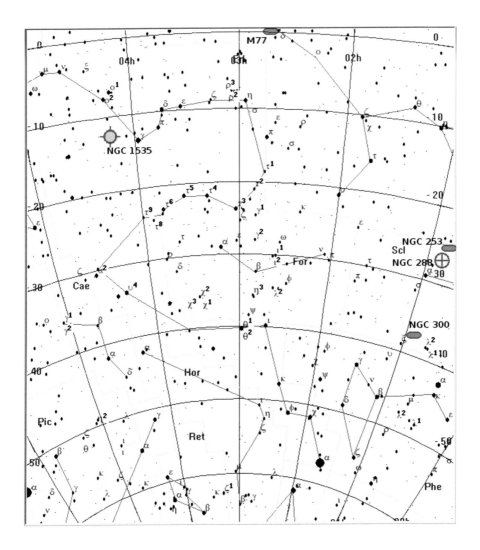

North RA 04 h 30 m to 07 h 30 m

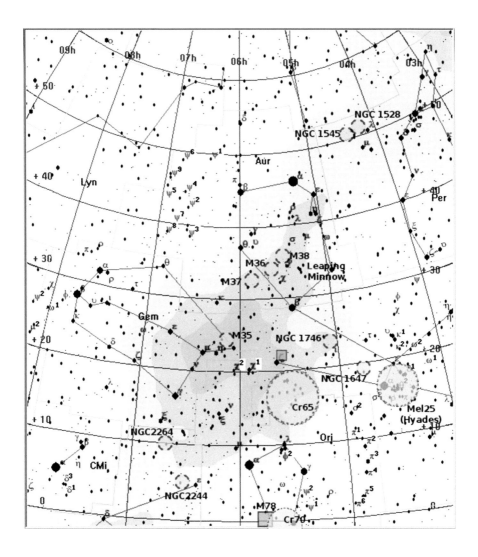

South RA 04 h 30 m to 07 h 30 m

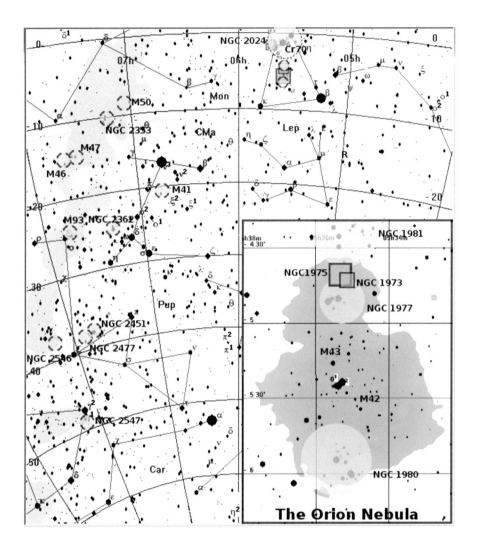

North RA 07 h 30 m to 10 h 30 m

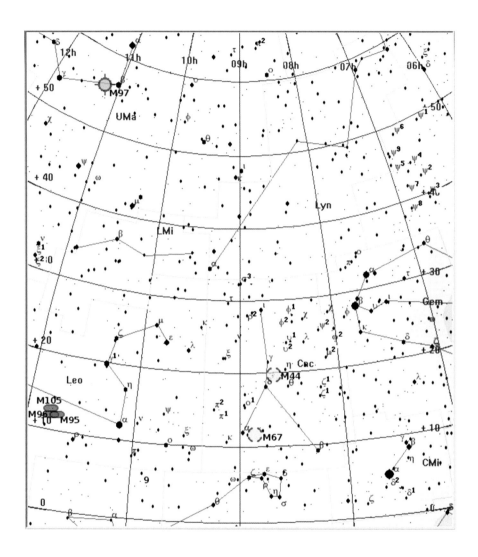

South RA 07 h 30 m to 10 h 30 m

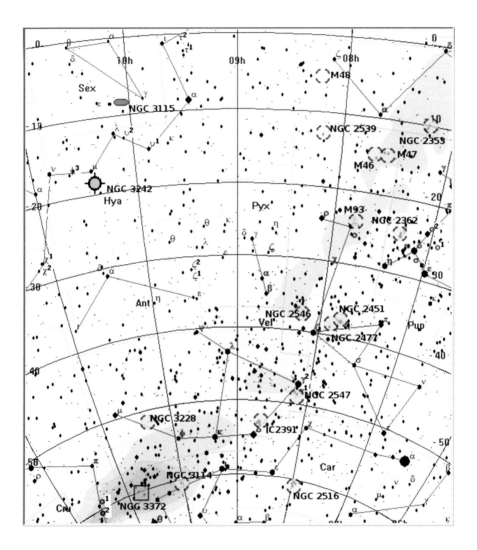

North RA 10 h 30 m to 13 h 30 m

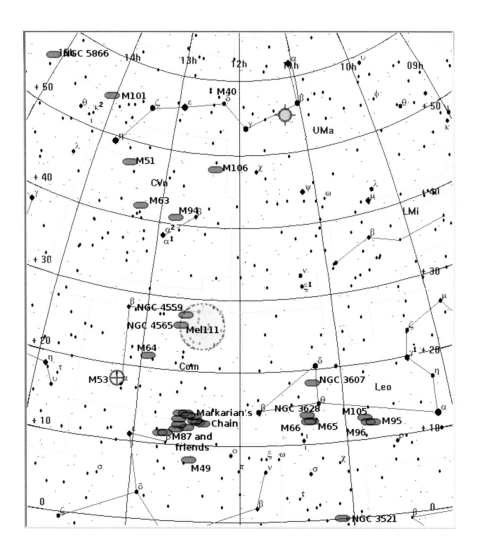

South RA 10 h 30 m to 13 h 30 m

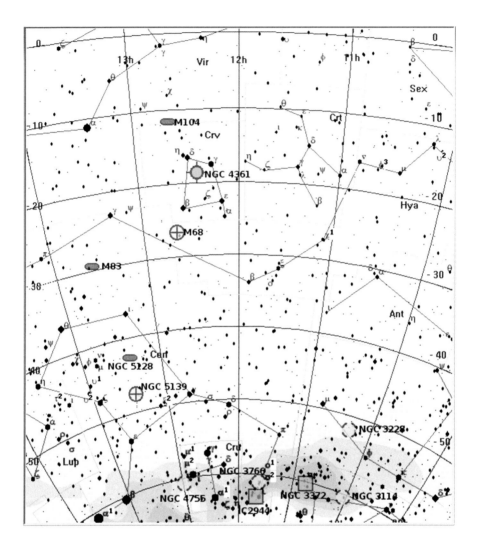

North RA 13 h 30 m to 16 h 30 m

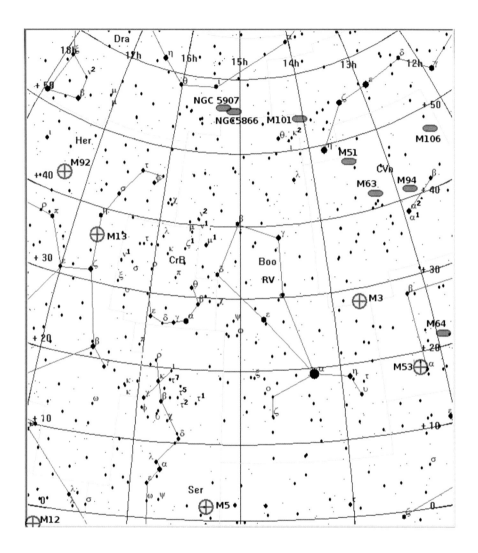

South RA 13 h 30 m to 16 h 30 m

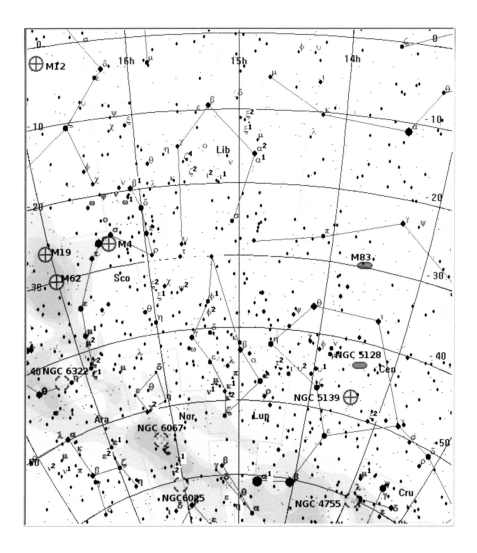

North RA 16 h 30 m to 19 h 30 m

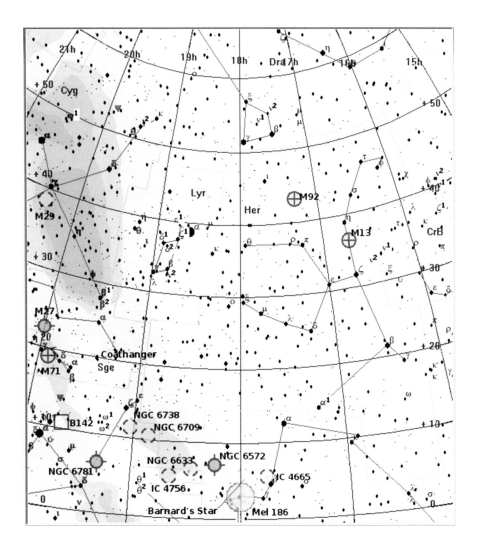

South RA 16 h 30 m to 19 h 30 m

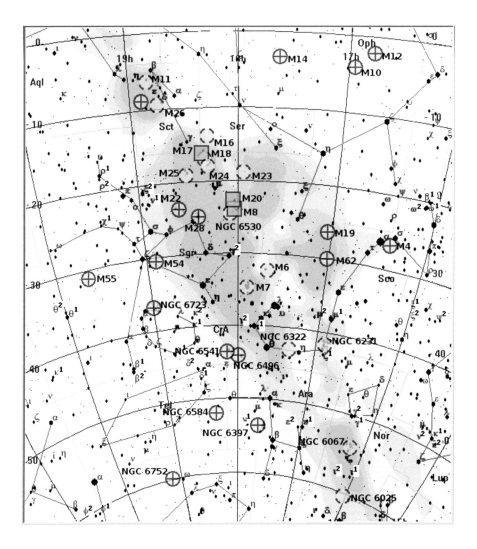

North RA 19 h 30 m to 22 h 30 m

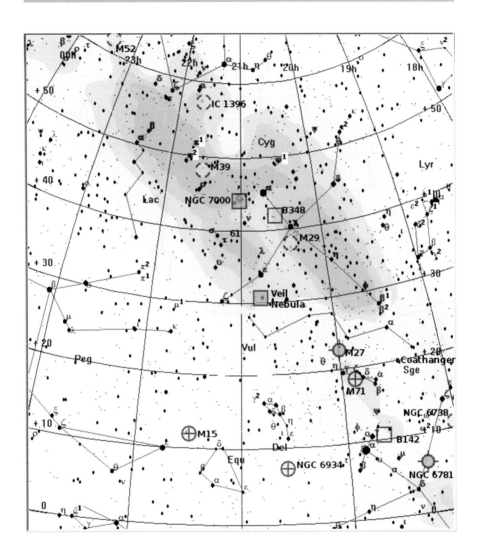

South RA 19 h 30 m to 22 h 30 m

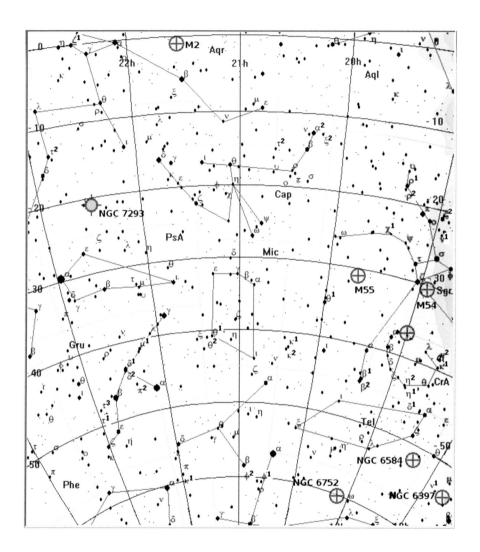

South Polar Region

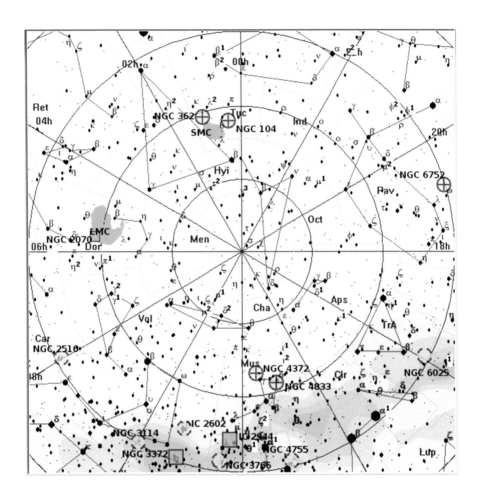

Objects by Type (Listed in Order of Right Ascension)

Asterisms

The Engagement Ring
Kemble's Cascade
The Leaping Minnow
M40
The Coathanger (Crr 399, Brocchi's Cluster, Al Sufi's Cluster)

Dark Nebulae

Barnard 142, 143 (Barnard's E)
LDN 906 (B 348, the Northern Coalsack)

Emission Nebulae

NGC 1499 (the California Nebula)
M78 (NGC 2068)
M42 (NGC 1976, the Great Orion Nebula)
M43 (NGC 1982)
NGC 2070 (Tarantula Nebula, Loop Nebula, 30 Doradus)
NGC 2024 (the Flame Nebula, the Burning Bush, the Ghost of Alnitak)
NGC 3372 (the η Carinae Nebula, the Homunculus Nebula)
M20 (NGC 6514, the Trifid Nebula)
M8 (NGC 6523, the Lagoon Nebula)
M17 (NGC 6618, the Omega Nebula or Swan Nebula)
NGC 7000 (the North American Nebula)

Galaxies

NGC 55
M31: the Great Andromeda Galaxy
NGC 247
NGC 253
NGC 292 (Small Magellanic Cloud)
NGC 300
M33 (NGC 598, the Pinwheel Galaxy)
M104 (NGC 4594, the Sombrero Galaxy)
M77 (NGC 1068)
NGC 1232
The Large Magellanic Cloud
NGC 2403
M81 (NGC 3031)
M82 (NGC 3034)
NGC 3115 (the Spindle Galaxy)
M95 (NGC 3351)
M96 (NGC 3368)
M105 (NGC 3379)
NGC 3521

NGC 3607
M65 (NGC 3623)
M66 (NGC 3627)
NGC 3628
M106 (NGC 4258)
M84 (NGC4374)
M86 (NGC4406)
Markarian's Chain
NGC 4438
NGC 4459
M49 (NGC 4472)
NGC 4473
NGC 4477
M87 (NGC 4486)
M88 (NGC 4501)
M91 (NGC 4501)
M89 (NGC 4552)
NGC 4559
NGC 4565 (Berenice's Hair Clip)
M90 (NGC 4569)
M58 (NGC 4579)
M59 (NGC 4621)
NGC 4631
M60 (NGC 4649)
NGC 4656
M94 (NGC 4736)
M64 (NGC 4826, the Black Eye Galaxy)
M63 (NGC 5055, the Sunflower Galaxy)
NGC 5128 (Centaurus A)
M51 (NGC 5194, the Whirlpool Galaxy)
M83 (NGC 5263)
M101 (NGC 5457)
NGC 5866 (M102)
NGC5907 (the Splinter Galaxy)

Globular Clusters

NGC 104 (47 Tucanae)
NGC 288
NGC 362
NGC 4372
M68 (NGC 4590)
NGC 4833

M53 (NGC 5024)
NGC 5139 (ω Centauri)
M3 (NGC 5272)
M5 (NGC 5904)
M4 (NGC 6121)
M13 (NGC 6205, the Great Hercules Globular Cluster)
M12 (NGC 6218)
M10 (NGC 6254)
M62 (NGC 6266)
M19 (NGC 6273)
M92 (NGC 6341)
M14 (NGC 6402)
NGC 6397
NGC 6496
NGC 6541
NGC 6584
M28 (NGC 6626)
M22 (NGC 6656)
M54 (NGC 6715)
NGC 6712
NGC 6723
NGC 6752
M55 (NGC 6809)
M71 (NGC 6838)
NGC 6934
M15 (NGC 7078)
M2 (NGC 7089)

Multiple Stars

ψ^1 Piscium
ζ Piscium
56 Andromedae
14 Arietis
θ Pictoris
σ Orionis
γ Leporis
9 Sextantis
δ Boötis
50 Boötis
ρ Ophiuchi
θ Serpentis
β Cygni (Albireo)

ε Sagittae
61 Cygni
Σ2809
ε Pegasi (Enif)

Open Clusters

NGC1746
NGC 3228
NGC 225
NGC 436
NGC 457 (the ET Cluster, the Owl Cluster)
NGC 654
NGC 663
Collinder 463
NGC 752
Stock 2 (the Muscleman Cluster)
NGC 884 and NGC 869 (the Perseus Double Cluster)
NGC 956
Melotte 15
M34 (NGC 1039)
NGC 1027
Stock 23
Melotte 20 (Collinder 39, the Alpha Persei Moving Cluster)
NGC 1342
M45 (the Pleiades)
NGC 1528
NGC 1545
Melotte 25 (the Hyades)
NGC 1647
Collinder 65
M38 (NGC 1912)
NGC 1981
NGC 1980
Collinder 70
M36 (NGC 1960)
M37 (NGC 2099)
M35 (NGC 2168)
NGC 2244
NGC 2264 (the Christmas Tree Cluster)
M41 (NGC 2287)
M50 (NGC 2323)
NGC 2353

NGC 2362
M47 (NGC 2422)
M46 (NGC 2437)
M93 (NGC 2447)
NGC 2451
NGC 2477
NGC 2516
NGC 2547
NGC 2539
NGC 2546
M48 (NGC 2548)
IC 2391 (the Omicron Velorum Cluster)
M44 (NGC 2632, Praesepe, the Beehive Cluster)
M67 (NGC 2682)
NGC 3114
IC 2602 (the Southern Pleiades)
NGC 3766
IC 2944; the Running Chicken (the λ Cen Nebula)
Melotte 111
NGC 4755 (the Jewel Box)
NGC 6025
NGC 6067
NGC 6231
NGC 6322
M6 (NGC 6405, the Butterfly Cluster)
IC 4665 (the Summer Beehive)
M7 (NGC 6475, Ptolemy's Cluster)
M23 (NGC 6494)
Melotte 186
NGC 6530
M24
NGC6603
M16 (NGC 6611, the Eagle Nebula)
M18 (NGC 6613)
NGC 6633
M25 (IC 4725)
IC 4756
M26 (NGC 6694)
M11 (NGC 6705, Wild Duck Cluster)
NGC6709
NGC 6738
M29 (NGC 6913)
M39 (NGC 7092)
IC1396
NGC 7209
NGC 7235

NGC 7243
NGC 7510
NGC 7686
Stock 12
NGC 7789
M52 (NGC7654)

Planetary Nebulae

NGC 1535
NGC 3242 (the Ghost of Jupiter)
M97 (NGC 3587, the Owl Nebula)
NGC 4361
NGC 6572
NGC 6781
M27 (NGC 6853, the Dumbbell Nebula, the Apple Core)
NGC 7293 (the Helix Nebula)

Reflection Nebulae

NGC 1973/5/7 (the Running Man)

Supernova Remnants

M1 (NGC 1952, the Crab Nebula)
Veil Nebula (NGC 6960, NGC 6992 & 6995)

Nearby Star

Barnard's Star

Variable Stars

o Ceti (Mira)
R Leporis (Hind's Crimson Star)
Y Canum Venaticorum (La Superba)

RV Boötis
U Sagittarii
μ Cephei (the Garnet Star)

Objects by Binocular Aperture (Listed in Order of Right Ascension)

The objects in this table are listed by the aperture of binocular for which I have prepared the finder chart. You should not take this to imply that this is necessarily the best size of binocular for an object; although in some instances this will be the case. It does imply that the object can be observed with a binocular of the size stated, although it may require specific conditions. Where this is the case, I have noted this in the object description.

In general, with the exception of objects that are too large to fit in the field of view, all objects will be better in larger-aperture binoculars.

50 mm

M31: the Great Andromeda Galaxy
NGC 292 (Small Magellanic Cloud)
M33 (NGC 598, the Pinwheel Galaxy)
NGC 663
14 Arietis
o Ceti (Mira)
NGC 884 and NGC 869 (the Perseus Double Cluster)
M34 (NGC 1039)
Melotte 20 (Collinder 39, the Alpha Persei Moving Cluster)
M45 (the Pleiades)
Kemble's Cascade
Melotte 25 (the Hyades)
The Leaping Minnow
Collinder 65
NGC 1981
M42 (NGC 1976, the Great Orion Nebula)
NGC 1980
M43 (NGC 1982)
Collinder 70
σ Orionis
γ Leporis
M35 (NGC 2168)
M41 (NGC 2287)
M50 (NGC 2323)
M47 (NGC 2422)
M46 (NGC 2437)
NGC 2451

Objects by Binocular Aperture (Listed in Order of Right Ascension) 177

IC 2391 (the Omicron Velorum Cluster)
M44 (NGC 2632, Praesepe, the Beehive Cluster)
NGC 3114
IC 2602 (the Southern Pleiades)
NGC 3372 (the η Carinae Nebula, the Homunculus Nebula)
Melotte 111
Y Canum Venaticorum (La Superba)
NGC 4755 (the Jewel Box)
NGC 5139 (ω Centauri)
M13 (NGC 6205, the Great Hercules Globular Cluster)
NGC 6231
M6 (NGC 6405, the Butterfly Cluster)
M7 (NGC 6475, Ptolemy's Cluster)
Melotte 186
M8 (NGC 6523, the Lagoon Nebula)
NGC 6530
M24
NGC 6603
IC 4756
M11 (NGC 6705, Wild Duck Cluster)
The Coathanger (Collinder 399, Brocchi's Cluster, Al Sufi's Cluster)
Albireo
M27 (NGC 6853, the Dumbbell Nebula, the Apple Core)
LDN 906 (B 348, the Northern Coalsack)
NGC 7000 (the North American Nebula)
M15 (NGC 7078)
M2 (NGC 7089)
IC 1396
μ Cephei (the Garnet Star)
ε Pegasi (Enif)

70 mm

NGC 225
NGC 654
Cr 463
Stock 2 (the Muscleman Cluster)
The Engagement Ring
Mel 15
NGC 1027
St 23
NGC 1342
NGC 1499 (the California Nebula)
NGC 1528
NGC 1647
R Leporis (Hind's Crimson Star)
NGC 1746

M38 (NGC 1912)
M36 (NGC 1960)
NGC 2024 (the Flame Nebula, the Burning Bush, the Ghost of Alnitak)
M78 (NGC 2068)
M37 (NGC 2099)
NGC 2244
NGC 2264 (the Christmas Tree Cluster)
M93 (NGC 2447)
M48 (NGC 2548)
M67 (NGC 2682)
M49 (NGC 4472)
M87 (NGC 4486)
M89 (NGC 4552)
M59 (NGC 4621)
M60 (NGC 4649)
M94 (NGC 4736)
M64 (NGC 4826, the Black Eye Galaxy)
M63 (NGC 5055, the Sunflower Galaxy)
M3 (NGC 5272)
M12 (NGC 6218)
M10 (NGC 6254)
M19 (NGC 6273)
M14 (NGC 6402)
IC 4665 (the Summer Beehive)
M23 (NGC 6494)
Barnard's Star
M28 (NGC 6626)
M22 (NGC 6656)
M26 (NGC 6694)
M29 (NGC 6913)
M39 (NGC 7092)
NGC 7209
NGC 7235
NGC 7243
NGC 7510
NGC 7686
St 12
NGC 7789

100 mm

NGC 55
NGC 3228
NGC 104 (47 Tucanae)
NGC 247
NGC 253
NGC 288

NGC 300
NGC 362
ψ^1 Piscium
ζ Piscium
NGC 436
NGC 457 (the ET Cluster, the Owl Cluster)
56 And
NGC 752
NGC 956
M104 (NGC 4594, the Sombrero Galaxy)
M77 (NGC 1068)
NGC 1232
NGC 1535
NGC 1545
The Large Magellanic Cloud
θ Pictoris
M1 (NGC 1952, the Crab Nebula)
NGC 1973/5/7 (the Running Man)
NGC 2070 (Tarantula Nebula, Loop Nebula, 30 Doradus)
NGC 2353
NGC 2362
NGC 2403
NGC 2477
NGC 2516
NGC 2547
NGC 2539
NGC 2546
9 Sextantis
M81 (NGC 3031)
M82 (NGC 3034)
NGC 3115 (the Spindle Galaxy)
NGC 3242 (the Ghost of Jupiter)
M95 (NGC 3351)
M96 (NGC 3368)
M105 (NGC 3379)
NGC 3521
M97 (NGC 3587, the Owl Nebula)
NGC 3607
M65 (NGC 3623)
M66 (NGC 3627)
NGC 3628
NGC 3766
IC 2944; the Running Chicken (the λ Cen Nebula)
M106 (NGC 4258)
M40

NGC 4361
M84 (NGC4374)
NGC 4372
M86 (NGC4406)
Markarian's Chain
NGC 4438
NGC 4459
NGC 4473
NGC 4477
M88 (NGC 4501)
M91 (NGC 4501)
NGC 4559
NGC 4565 (Berenice's Hair Clip)
M90 (NGC 4569)
M58 (NGC 4579)
M68 (NGC 4590)
NGC 4631
NGC 4656
NGC 4833
M53 (NGC 5024)
NGC 5128 (Centaurus A)
M51 (NGC 5194, the Whirlpool Galaxy)
M83 (NGC 5263)
M101 (NGC 5457)
RV Boötis
NGC 5866 (M102)
δ Boötis and 50 Boötis
NGC5907 (the Splinter Galaxy)
M5 (NGC 5904)
NGC 6025
NGC 6067
M4 (NGC 6121)
ρ Ophiuchi
M62 (NGC 6266)
M92 (NGC 6341)
NGC 6322
NGC 6397
NGC 6496
M20 (NGC 6514, the Trifid Nebula)
NGC 6541
NGC 6572
NGC 6584
M16 (NGC 6611, the Eagle Nebula)
M18 (NGC 6613)
M17 (NGC 6618, the Omega Nebula or Swan Nebula)
NGC 6633

M25 (IC 4725)
U Sagittarii
NGC6709
NGC 6712
M54 (NGC 6715)
θ Serpentis
NGC 6723
NGC 6738
NGC 6752
NGC 6781
ε Sagittae
Barnard 142, 143 (Barnard's E)
M55 (NGC 6809)
NGC 7293 (the Helix Nebula)
M52 (NGC7654)

Objects by Constellation

Andromeda

M31: the Great Andromeda Galaxy
56 And
NGC 752
NGC 956
NGC 7686

Aquarius

M2 (NGC 7089)
Σ2809
NGC 7293 (the Helix Nebula)

Aquila

NGC6709
NGC 6738
NGC 6781
Barnard 142, 143 (Barnard's E)

Ara

NGC 6397

Aries

14 Arietis

Auriga

The Leaping Minnow
M38 (NGC 1912)
M36 (NGC 1960)
M37 (NGC 2099)

Boötes

RV Boötis
δ Boötis and 50 Boötis

Camelopardalis

St 23
Kemble's Cascade
NGC 2403

Cancer

M44 (NGC 2632, Praesepe, the Beehive Cluster)
M67 (NGC 2682)

Canis Major

M41 (NGC 2287)
NGC 2362

Carina

NGC 2516
NGC 3114
IC 2602 (the Southern Pleiades)
NGC 3372 (the η Carinae Nebula, the Homunculus Nebula)

Cassiopeia

NGC 225
NGC 436
NGC 457 (the ET Cluster, the Owl Cluster)
NGC 654
NGC 663
Cr 463
Stock 2 (the Muscleman Cluster)
Mel 15
NGC 1027
M52 (NGC7654)
St 12
NGC 7789

Centaurus

NGC 3766
IC 2944; the Running Chicken (the λ Cen Nebula)
NGC 5128 (Centaurus A)
NGC 5139 (ω Centauri)

Cepheus

IC1396
μ Cephei (the Garnet Star)
NGC 7235
NGC 7510

Cetus

NGC 247
o Ceti (Mira)
M77 (NGC 1068)

Coma

Melotte 111
M88 (NGC 4501)
M91 (NGC 4501)
NGC 4559
NGC 4565 (Berenice's Hair Clip)
M64 (NGC 4826, the Black Eye galaxy)
M53 (NGC 5024)

Corona Australis

NGC 6496
NGC 6541

Corvus

NGC 4361

Crux

NGC 4755 (the Jewel Box)

Canes Venatici

M106 (NGC 4258)
NGC 4631
NGC 4656
Y Canum Venaticorum (La Superba)
M94 (NGC 4736)
M63 (NGC 5055, the Sunflower Galaxy)
M51 (NGC 5194, the Whirlpool Galaxy)
M3 (NGC 5272)

Cygnus

Albireo
M29 (NGC 6913)
LDN 906 (B 348, the Northern Coalsack)
Veil Nebula (NGC 6960, NGC 6992 & 6995)
NGC 7000 (the North American Nebula)
61 Cygni
M39 (NGC 7092)

Delphinus

NGC 6934

Dorado

The Large Magellanic Cloud
NGC 2070 (Tarantula Nebula, Loop Nebula, 30 Doradus)

Draco

NGC 5866 (M102)
NGC5907 (the Splinter Galaxy)

Eridanus

NGC 1232
NGC 1535

Gemini

M35 (NGC 2168)

Hercules

M13 (NGC 6205, the Great Hercules Globular Cluster)
M92 (NGC 6341)

Hydra

M48 (NGC 2548)
NGC 3242 (the Ghost of Jupiter)
M68 (NGC 4590)
M83 (NGC 5263)

Lacerta

NGC 7209
NGC 7243

Leo

M95 (NGC 3351)
M96 (NGC 3368)

M105 (NGC 3379)
NGC 3521
NGC 3607
M65 (NGC 3623)
M66 (NGC 3627)
NGC 3628

Lepus

R Leporis (Hind's Crimson Star)
γ Leporis

Monoceros

NGC 2244
NGC 2264 (the Christmas Tree Cluster)
M50 (NGC 2323)
NGC 2353
NGC 4372
NGC 4833

Norma

NGC 6067

Ophiuchus

ρ Ophiuchi
M12 (NGC 6218)
M10 (NGC 6254)
M62 (NGC 6266)
M19 (NGC 6273)
M14 (NGC 6402)
IC 4665 (the Summer Beehive)
Barnard's Star
Melotte 186
NGC 6572
NGC 6633

Orion

Collinder 65
NGC 1973/5/7 (the Running Man)
NGC 1981
M42 (NGC 1976, the Great Orion Nebula)
NGC 1980
M43 (NGC 1982)
Collinder 70
σ Orionis
NGC 2024 (the Flame Nebula, the Burning Bush, the Ghost of Alnitak)
M78 (NGC 2068)

Pavo

NGC 6752

Pegasus

M15 (NGC 7078)
ε Pegasi (Enif)

Perseus

NGC 884 and NGC 869 (the Perseus Double Cluster)
M34 (NGC 1039)
Melotte 20 (Collinder 39, the Alpha Persei Moving Cluster)
NGC 1342
NGC 1499 (the California Nebula)
NGC 1528
NGC 1545

Pictor

θ Pictoris

Pisces

ψ^1 Piscium
ζ Piscium

Puppis

M47 (NGC 2422)
M46 (NGC 2437)
M93 (NGC 2447)
NGC 2451
NGC 2477
NGC 2539
NGC 2546

Sagitta

ε Sagittae
M71 (NGC 6838)

Sagittarius

M23 (NGC 6494)
M20 (NGC 6514, the Trifid Nebula)
M8 (NGC 6523, the Lagoon Nebula)
NGC 6530
M24
NGC6603
M18 (NGC 6613)
M17 (NGC 6618, the Omega Nebula or Swan Nebula)
M28 (NGC 6626)
M25 (IC 4725)
U Sagittarii
M22 (NGC 6656)
M54 (NGC 6715)
NGC 6723
M55 (NGC 6809)

Scorpius

M4 (NGC 6121)
NGC 6231

NGC 6322
M6 (NGC 6405, the Butterfly Cluster)
M7 (NGC 6475, Ptolemy's Cluster)

Sculptor

NGC 55
NGC 253
NGC 288
NGC 300

Scutum

M26 (NGC 6694)
M11 (NGC 6705, Wild Duck Cluster)
NGC 6712

Serpens

M5 (NGC 5904)
M16 (NGC 6611, the Eagle Nebula)
IC 4756
θ Serpentis

Sextans

9 Sextantis
NGC 3115 (the Spindle Galaxy)

Taurus

M45 (the Pleiades)
Melotte 25 (the Hyades)

NGC 1647
NGC 1746
M1 (NGC 1952, the Crab Nebula)

Telescopium

NGC 6584

Triangulum

M33 (NGC 598, the Pinwheel Galaxy)

Triangulum Australis

NGC 6025

Tucana

NGC 104 (47 Tucanae)
NGC 292 (Small Magellanic Cloud)
NGC 362

Ursa Major

M81 (NGC 3031)
M82 (NGC 3034)
M97 (NGC 3587, the Owl Nebula)
M40
M101 (NGC 5457)

Ursa Minor

The Engagement Ring

Vela

NGC 3228
NGC 2547
IC 2391 (the Omicron Velorum Cluster)

Virgo

M104 (NGC 4594, the Sombrero Galaxy)
M84 (NGC4374)
M86 (NGC4406)
Markarian's Chain
NGC 4438
NGC 4459
M49 (NGC 4472)
NGC 4473
NGC 4477
M87 (NGC 4486)
M89 (NGC 4552)
M90 (NGC 4569)
M58 (NGC 4579)
M59 (NGC 4621)
M60 (NGC 4649)

Vulpecula

The Coathanger (Collinder 399, Brocchi's Cluster, Al Sufi's Cluster)
M27 (NGC 6853, the Dumbbell Nebula, the Apple Core)

Bibliography

Burnham, R., ***Burnham's Celestial Handbook Vol 1***, New York, Dover Publications Inc., 1978, ISBN 0-486-23567-X
Moore, P., ***Exploring the Night Sky with Binoculars***, Cambridge, Cambridge University Press, 1986, ISBN 0521368669
Moore, P. (ed.), ***Philip's Astronomy Encyclopedia***, London, Philip's, 2002, ISBN 0001032086

Chapter 11

December Solstice to March Equinox (RA 04:00 h to 10:00 h)

Perseus: Emission Nebula: NGC 1499 (the *California Nebula*) (70 mm)

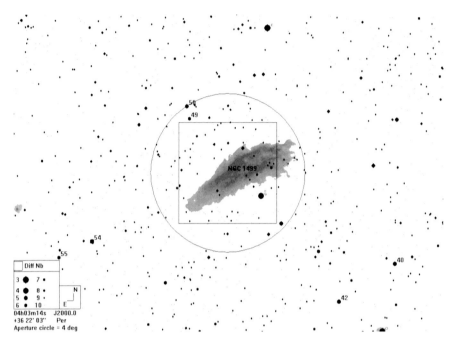

The *California Nebula* is less than a degree N of ξ **Per**.

A 15×70 is the ideal binocular, for this is a large, very faint (it has an integrated magnitude of 6.0, but this is spread over more than a square degree) nebula, which requires excellent conditions to become visible—under these conditions, some observers are able to see it with the unaided eye. With a dark, transparent sky and averted vision, this accumulation of gas, which is energized by ultraviolet radiation from the runaway star, ξ Per, becomes faintly and eerily apparent, usually starting at the SE region, then gradually extending northwards as you are able to see more of it.

If you have a Hβ filter, this can make it much easier to see if you hold it in front of an eyepiece. A UHC can also help, but an [O-III] makes it invisible!

Perseus: Open Cluster: NGC 1528 (70 mm)

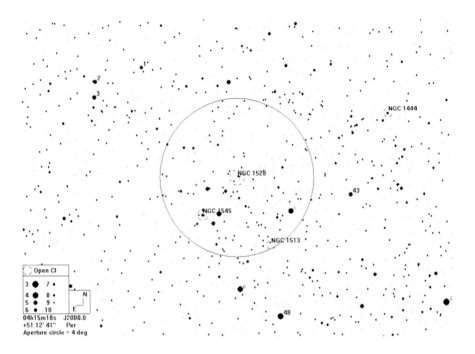

From δ **Per**, scan 2 fields to the NE to find λ **Per**. Place λ **Per** near the SW of the field and **NGC 1528** will appear as a misty patch to the NE.

Only a few stars are resolved in this bright cluster, which still appears mostly as a misty patch even in big binoculars. It is one of several objects that could easily have been in Messier's catalogue of comet-like objects. Also look at **NGC 1545** to the SW, which can fit into the same field. By comparison, this is a rather poor cluster, being sparser and smaller.

Eridanus: Planetary Nebula: NGC 1535 (100 mm)

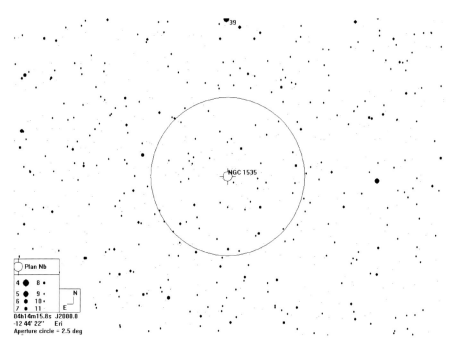

NGC 1535 is located 4° ENE of **γ Eri**.

This small, 10th magnitude planetary nebula is a moderately easy object at 37×100 as long as the sky is transparent and free of light pollution. However, this magnification shows no structure.

Taurus: Open Cluster: Melotte 25 (C41, the *Hyades*) (50 mm)

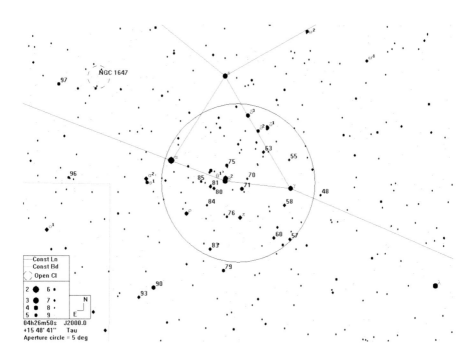

The Hyades is the large cluster adjacent to α **Tauri** (*Aldebaran*), which is not itself a member.

This cluster, the second closest to us at a distance of about 150 light-years, overflows a 5° field of view. For this reason, it is far better in binoculars than it is in a telescope. The brighter stars form the "V" shape with which we are familiar as the head of the bull. The Hyades lies at the approximate center of a larger grouping of stars, the Taurus Moving Cluster, some members of which are over 45° from the Hyades. Several tens of stars are revealed by 10×50 binoculars. Also worth a look is the cluster NGC 1647 which is just to the NW.

Taurus: Open Cluster: NGC1647 (70 mm)

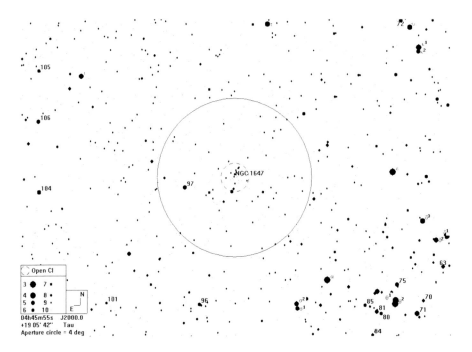

NGC 1647 is one field NE of *Aldebaran* (α **Tau**).

This is a cluster that deserves to be far better known. It is under-observed because of its proximity to its illustrious neighbors, the *Pleiades* and the *Hyades*. This big, somewhat sparse, grouping of stars is much better in binoculars than it is in a telescope, where it does not always appear to be an obvious cluster.

Taurus: Open Cluster: NGC 1746 (70 mm)

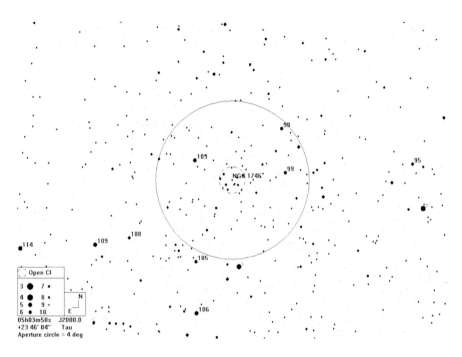

NGC 1746 is two fields SW of *El Nath* (β **Tau**), halfway between the star and **NGC 1647**.

This beautiful cluster is much looser cluster than the nearby **NGC1647**. Nineteen stars are resolved with direct vision.

Taurus: Supernova Remnant: M1 (NGC 1952, the Crab Nebula) (100 mm)

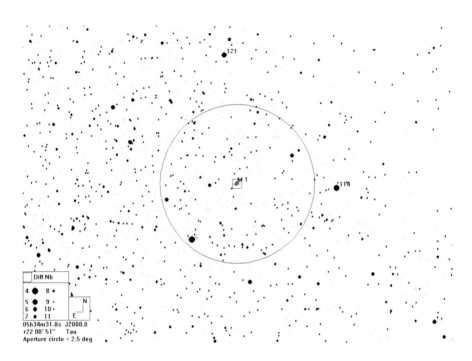

Place ζ **Tau** at the SE edge of the field, and **M1** will appear in the middle.

In the year 1758, on the night of August 28, a young assistant at the Naval Observatory at the Hotel de Cluny in Paris discovered what appeared to be a comet in the constellation of Taurus. This young man had been charged by the observatory director (Joseph-Nicolas Delisle) to find Halley's Comet, the return of which had been predicted for that year. The assistant was unable to observe again for 2 weeks and, when he did, his new "comet" had not moved. This object in Taurus became the first object in the young Charles Messier's catalogue "fuzzy blobs" that should not be mistaken for comets, and thus he sowed the seeds for many sleepless nights, around the end of March and beginning of April, for amateurs who attempt his eponymous "marathon" of observing the entire catalogue between dusk and dawn.

The object in Taurus was later found to be the remnant of a supernova of 1054 that was visible for 2 years and was even briefly a daylight object.

For all that illustrious past, **M1** is a fairly boring object to observe with binoculars; it shows as nothing except a small fuzzy patch which is difficult to see unless the sky is very dark.

Lepus: Variable Star: R Leporis (*Hind's Crimson Star*) (70 mm)

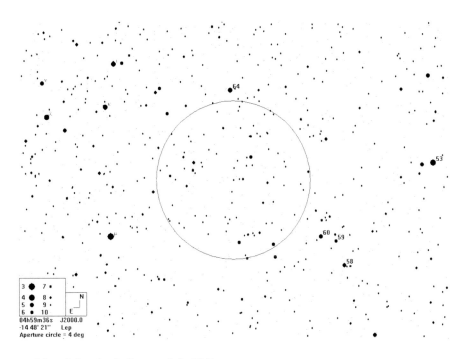

Mean Magnitude Range: 5.9–10.5
Mean Period: 427 days

Follow the line from α **Lep** to μ **Lep** a further 1½ fields, where **R Lep** should be visible and identifiable by its color.

R Lep is a Mira-type variable, but is not included for this reason, but because of its color. It is a candidate for the reddest visual star, hence its common name. Its color is obviously most impressive when it is near maximum.

Lepus: Double Star: γ Leporis (50 mm)

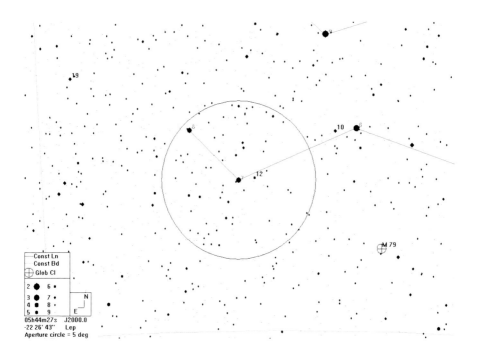

The star is visible to the naked eye.

The members of **γ Leporis** are separated by just over 1½ arcmin. In 10×50 binoculars, the 6th magnitude fainter member is noticeably more orange (spectral type G) than the yellow (spectral type F) 3.6th magnitude primary. Also seek out M79 (NGC 1904) that is a 5° field to the SE; it appears as a fuzzy star.

Auriga: Asterism: The *Leaping Minnow* (50 mm)

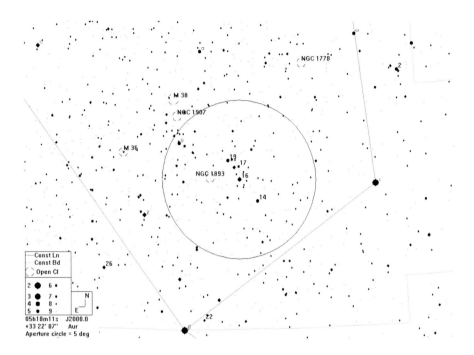

The *Leaping Minnow* asterism is an informal grouping of stars that includes **14**, **16**, **17**, and **19 Aurigae**.

If you include the water "splash," there are 30 or more stars of 9th magnitude and above, with the minnow itself being defined by half a dozen 5th and 6th magnitude stars. Presumably, the asterism gets its name from the similarity of the pattern of its bright stars to those of *Delphinus*.

Auriga: Three Open Clusters: M36 (NGC 1960), M37 (NGC 2099), and M38 (NGC 1912) (70 mm)

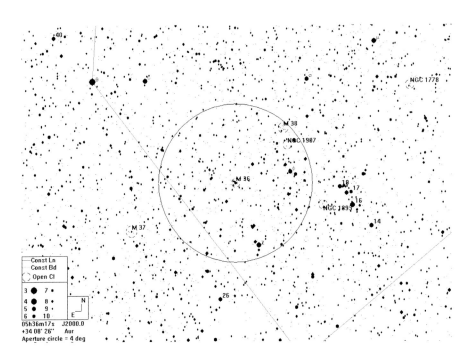

Midway between β **Tauri** (*El Nath*) and θ **Aurigae** is a slightly curved chain of three, evenly spaced, 7th magnitude stars. With this chain at the center of the field, M36 lies to the NW and M37 lies to the SE, both within the same 4° field. With M36 at the center of the field, M38 is just inside the NW edge.

The 6th magnitude M36 is approximately round and about a third of the diameter of the Moon. Depending on the sky conditions and the quality of the binoculars, you may be able to resolve up to about half a dozen of the brightest stars.

M37 is about twice as large as M36 and brighter overall (magnitude 5.6), as a consequence of having many more stars, but the individual stars themselves are fainter.

M38 (magnitude 6.4) is intermediate in size between M36 and M37. In good conditions, 15×70 binoculars may resolve over a dozen stars.

Dorado: Galaxy and Emission Nebula: *Large Magellanic Cloud* and NGC 2070 (C103, *Tarantula Nebula, Loop Nebula*, 30 Doradus) (100 mm)

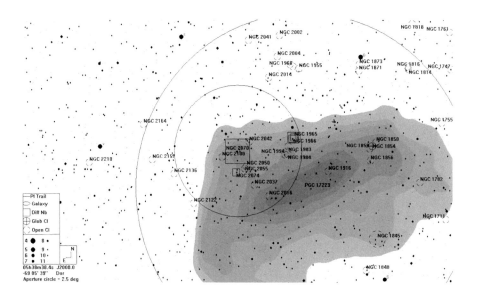

The chart is centered on **NGC 2070**.

The ***Large Magellanic Cloud*** (***LMC***) is easily visible to the naked eye. The ***Tarantula*** is situated within the ***LMC*** and makes an approximate equilateral triangle with ν and ε **Dor**.

Binoculars of any size give a breathtaking view of the ***LMC***, which is the brightest of our companion galaxies.

The ***Tarantula*** is very bright, being distinguishable to the naked eye if sky conditions permit. It is the largest known emission nebula and, if it were situated at the same distance as **M42** in Orion, it would be sufficiently bright to cast shadows.[1] The structure, some of which is visible in binoculars, gives it its common names, but it requires a larger aperture and more magnification to reveal significant detail.

---

[1] Moore 1986, p. 96

Pictor: Double Star: θ Pictoris (100 mm)

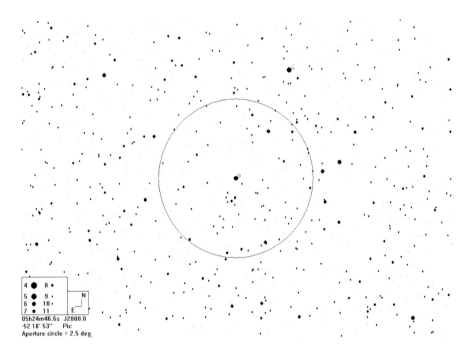

θ **Pic** makes the right angle of the right triangle that has β and ζ **Pic** at its other apexes.

Pictor is a constellation that is unremarkable to the naked eye, but which comes into its own with binoculars. θ **Pic** is a pair of almost equal brilliant white stars (just fainter than 6th magnitude) that is separated by 38 arcsec.

Orion: Open Cluster: Collinder 65 (50 mm)

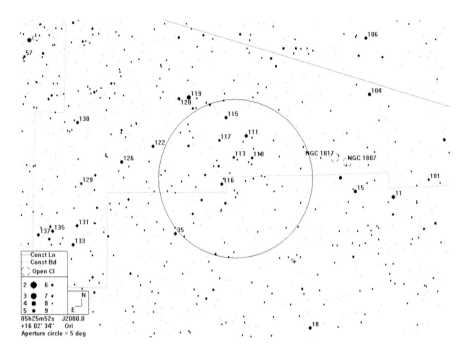

Cr65 is 6½° to the NNW of *Meissa* (λ **Ori**). α, λ, and γ **Ori** make an arrowhead that points to the cluster which, at magnitude 3.3, is visible to the unaided eye as a misty patch of light.

With a diameter of nearly 4°, Cr65 is an ideal object for 50 mm binoculars, which reveal over 50 stars in this sparse cluster.

Orion: Nebulosity and Clusters: M42 (NGC 1976), M43 (NGC 1982), NGC 1973, 1975, 1977, and 1980 (50 mm)

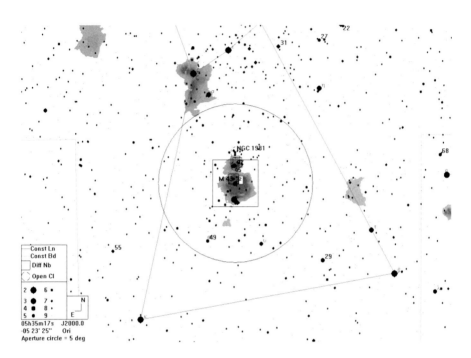

The *Great Nebula* in Orion is visible to the naked eye as the fuzzy middle star of the sword.

M42 and the connected **M43** benefit greatly from mounted binoculars. Even in small binoculars, a wealth of detail becomes apparent, especially if you use averted vision. Give it time: the longer you look, the more you see of the nebulosity and the cluster of stars whose light it reflects. If you have larger binoculars of higher magnification, see if you can resolve the *Trapezium* (θ **Orionis**) into separate stars. Binoculars will show that the other two "stars" of the sword into are also clusters. The northern one still has some reflection nebulosity (**NGC 1973, 1975, and 1977**) that may hint of its presence on dark, transparent nights, but the older southern cluster (**NGC 1980**) has none at all. These two clusters, which are older than that associated with the star-birth region of the *Great Nebula*, indicate the fate of the *Great Nebula* itself.

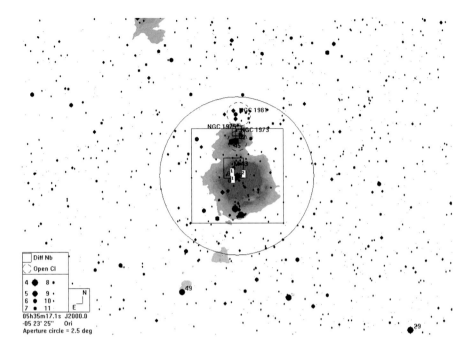

This naked-eye object is found in the middle of Orion's sword.

The *Great Nebula* is listed separately among the objects for 50-mm binoculars, but double the aperture and triple or quadruple the magnification, and it is almost like looking at a different object, especially on a transparent dark night. In big binoculars, far more fine detail becomes visible; it seems that the longer you look, the more you see, and a false stereopsis emerges. Look for structure around the "fish mouth" and in the "wings." The Trapezium (θ **Ori**) becomes resolvable and can be resolved into four stars with good optics and steady seeing.

This showpiece of the northern winter skies is one to be enjoyed over and over again.

Orion: Open Cluster: Cr 70 (50 mm)

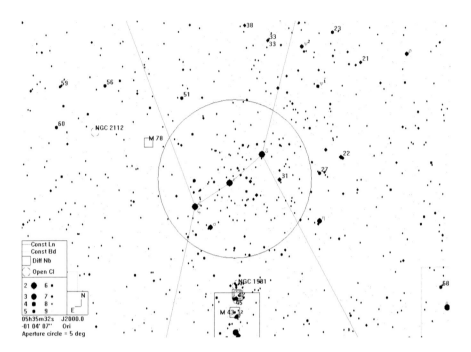

Cr 70 is the cluster that almost everybody has seen and yet nobody knows! It is the cluster that includes the stars of Orion's Belt and **σ Ori** (above). 10×50 binoculars will reveal many tens of stars in the 3° expanse of the cluster, making this one of several objects that is significantly better in medium binoculars than in most telescopes.

Orion: Multiple Star: σ Orionis (50 mm)

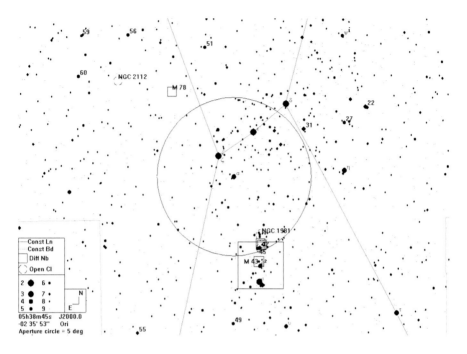

σ **Orionis** is visible to the naked eye as a 4th magnitude star about a degree to the SE of the easternmost belt star (ζ **Ori**, *Alnitak*).

σ **Ori** is, in fact, a multiple star consisting of six components. You should easily be able to split it into two components with a 10×50 binocular and see the blue 6th magnitude E component that is 43 arcsec from the white primary, but more magnification will be needed to separate out the two additional components. The final visual split requires the high power of a very good telescope. The presence of a 6th member is inferred from spectroscopic radial velocity measurements.

Orion: Nebula: NGC 2024 (*the Flame Nebula, the Burning Bush, the Ghost of Alnitak*) (70 mm)

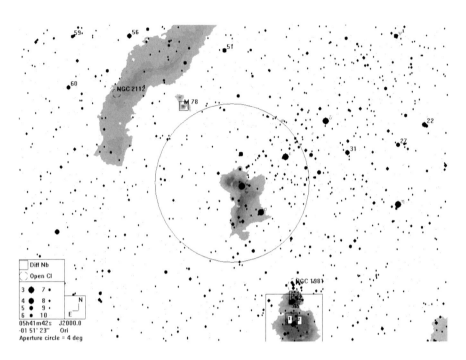

NGC2024 extends NE from *Alnitak* (ζ **Ori**).

The first time I tried to see the **Flame Nebula** in binoculars, it was unexpectedly easy to detect the dark dust lane through the middle of it, once Alnitak was just outside the field of view. This does, of course, require binoculars that have good control of stray light. Once you have seen this dark lane, you should be able to detect the shape of the rest of the nebula by using, initially at least, averted vision.

Orion: Emission Nebula: M78 (NGC 2068) (70 mm)

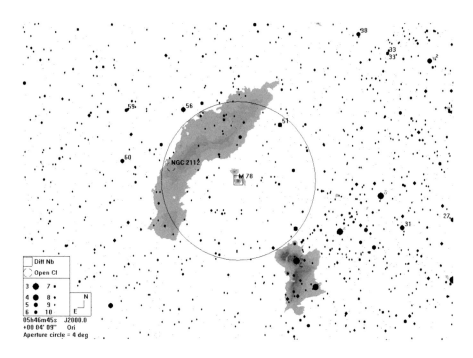

M78 is 2½° to the NNE of ***Alnitak*** (**ζ Ori**), and can therefore be found in the same field as the star.

Under very transparent skies, I have seen the 8th magnitude **M78** with 42-mm binoculars, but I find that 70 mm is required to bring out the abruptness in the way the northern edge darkens (dust lane?). This gives the southern region the appearance of a comet tail extending away from a northern coma, serving as a reminder of why Charles Messier drew up his catalogue of objects that should not be confused with comets.

Also take time with the two 10th magnitude stars that lie just to the north: the closer of the two has associated nebulosity (**NGC 2071**), which I have seen with 100 mm and averted vision under UK skies, but not 70 mm, although I see no reason why it should not be visible with the smaller binocular under very dark skies.

Gemini: Open Cluster: M35 (NGC 2168) (50 mm)

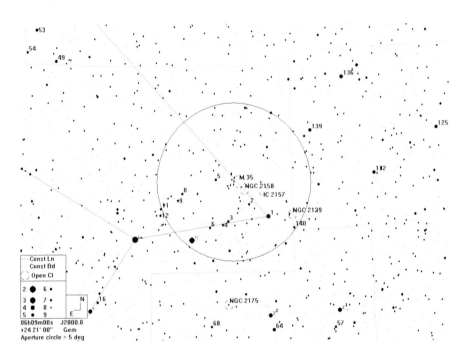

If you place η **Geminorum** at the SE edge of a 5° field, M35 will be in the center.

M35 deserves its nickname, the *Queen of Clusters*. It is a superb open cluster, about the size of the Moon and consisting of over 300 stars, of which 15 or so are resolvable in 10×50 binoculars. Using averted vision if necessary, see if you can glimpse two other open clusters, NGC 2158, which is half a degree to the SE, and the slightly more difficult IC 2157, which is a degree to the ESE.

Monoceros: Open Cluster: NGC 2239 (NGC 2244, C50) (70 mm)

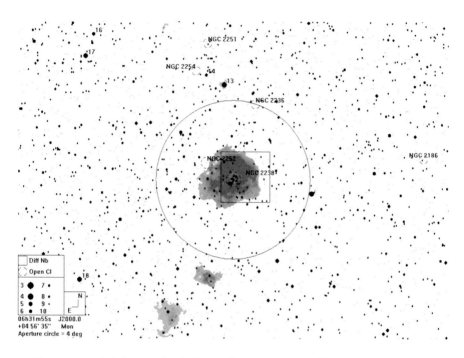

There are no bright stars in the immediate region. Possibly the simplest star hop is to follow the line from λ **Ori** through *Betelgeuse* (α **Ori**) for a further three 2.5° fields until you come to ε **Mon**. Just under one field E of ε **Mon**, **NGC2239** (also designated **NGC 2244**) is the cluster of stars that appears around **12 Mon**, which is a foreground object, not a member of the cluster.

Although **NGC 2239** is often given in lists for 50 mm binoculars, it comes into its own with larger glasses, where the yellow-orange **12 Mon** stands out against the predominantly blue-white stars of the cluster. Under ideal conditions, and with averted vision, it is possible to glimpse the surrounding Rosette Nebula as a slight brightening of the sky background.

Monoceros: Open Cluster: NGC 2264 (the *Christmas Tree Cluster*) (70 mm)

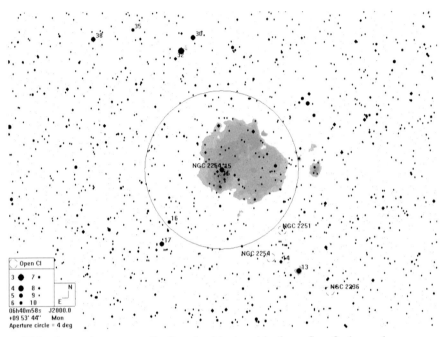

This cluster is most easily found with the aid of a reflex finder or low-power wide-field binoculars. Find the center of a line joining ***Betelgeuse*** (α **Ori**) to β **CMi** and offset 2° in the direction of γ **Gem** (*Alhena*). The cluster surrounds the star 15 Mon.

This bright cluster has a characteristic wedge shape from which it derives its common name. The paucity of faint stars is thought to be due to a significant amount of interstellar dust in the region. The surrounding nebulosity of the *Cone Nebula* is not normally visible but can be teased out with the aid of a UHC filter.

Monoceros: Open Cluster: M50 (NGC 2323) (50 mm)

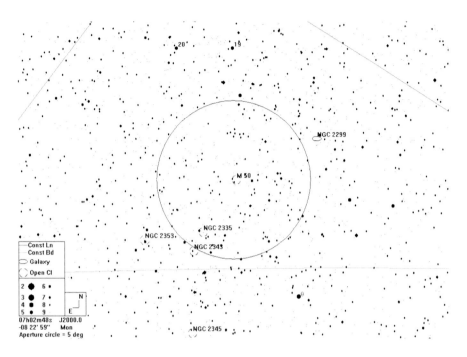

M50 is situated 4° to the NNW of the 4th magnitude **θ Canis Majoris**.

M50 is a bright open cluster in which 10×50 binoculars will resolve a few stars against the background glow of the hundred or so stars that comprise it.

Monoceros: Open Cluster: NGC 2353 (100 mm)

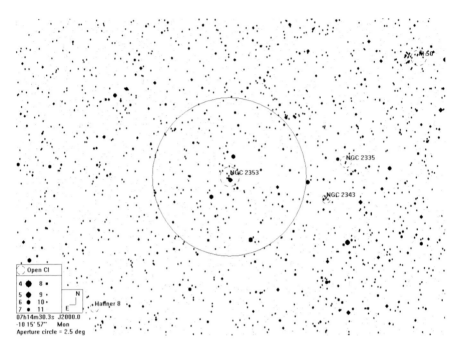

NGC 2353 can be found by hopping two and a half fields W of α **Mon**.

NGC 2353 is a particularly fine cluster in binoculars at ×37, in which several stars are resolved. It is not particularly dense, permitting the appreciation of the stars that become visible.

Canis Major: Open Cluster: M41 (NGC 2287) (50 mm)

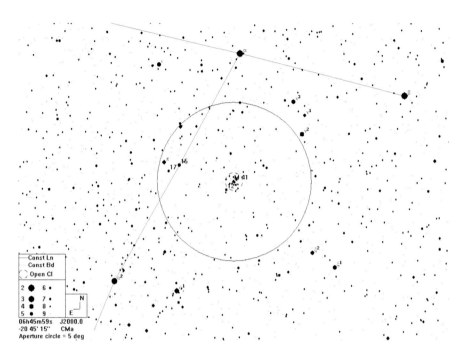

M41 is 4° south of α **Canis Majoris** (*Sirius*).

This cluster is visible to the naked eye (it was noted by Aristotle) in a transparent sky. In 10×50 binoculars, half a dozen or so stars are resolved against a background glow.

Canis Major: Open Cluster: NGC 2362 (C64) (100 mm)

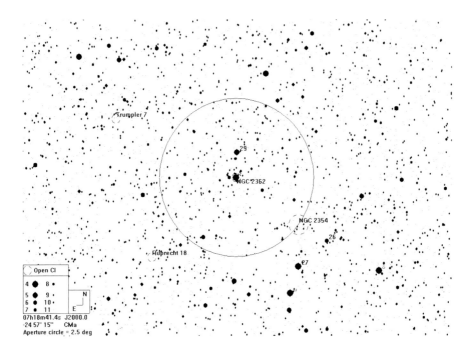

NGC 2362 surrounds the 4th magnitude τ **CMa**, which is just over one field to the NE of δ **CMa**.

This is a superb cluster in big binoculars, showing about 20 stars at ×37. It contains several very blue (spectral class O) stars; the brightest star, τ **CMa**, is a brilliant bluish white.

Puppis: Open Clusters: M46 (NGC 2437) and M47 (NGC 2422) (50 mm)

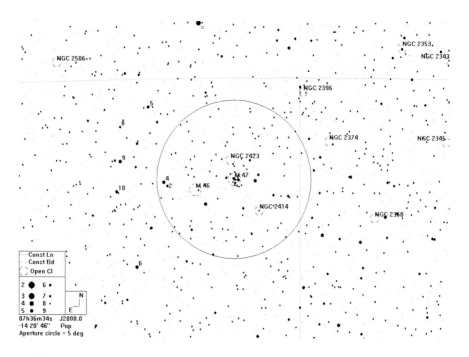

This pair of clusters lies 5° to the south of α **Monocerotis**; alternatively, if you cannot identify α **Mon**, you can find the clusters by panning two and a half 5° fields E and then half a 5° field N from α **CMa** (*Sirius*).

M46 and **M47** offer, in the same field of view, a comparison between the binocular appearances of open clusters. **M46** is far more compact and you may not be able to resolve any stars in 10×50 binoculars. On the other hand, **M47** is much looser and over half a dozen stars can be easily resolved with the 10×50s.

Camelopardalis: Galaxy: NGC 2403 (C7) (100 mm)

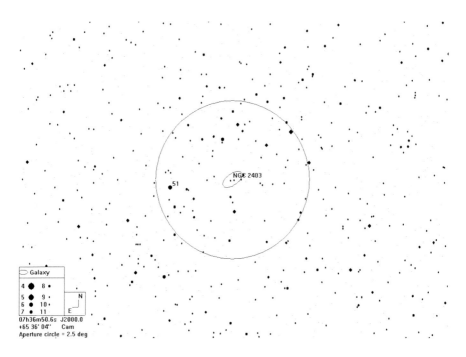

From *Mirak* (β **UMa**), hop 10°WSW to υ **UMa** and then a further 10° to o **UMa**. From o **UMa**, hop 7° to the 6th magnitude **51 UMa** and place it on the E edge of the field of view. **NGC 2403** will be near the center.

NGC 2403 is an obliquely viewed spiral galaxy. It will not show any spiral structure in big binoculars, but you should detect the brighter core. It looks oblate in adequate conditions and appears more elongated (2:1) in darker skies.

Carina: Open Cluster: NGC 2516 (C96) (100 mm)

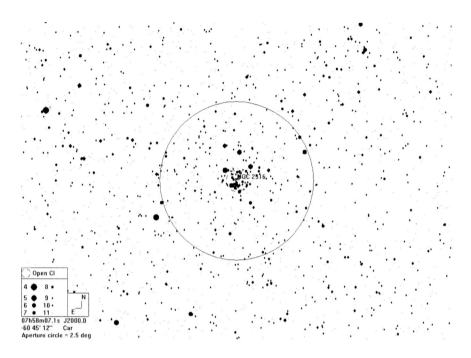

This easy naked-eye cluster is just over 3° 1½ fields SW of ε **Car** (*Evior*).

Although this superb cluster could easily have been included among the 50-mm objects, it is so much better at 37×100 than it is at 10×50 that I consider it more appropriately placed here. Expect to see over 30 stars if the conditions are right.

Vela: Open Cluster: NGC 2547 (100 mm)

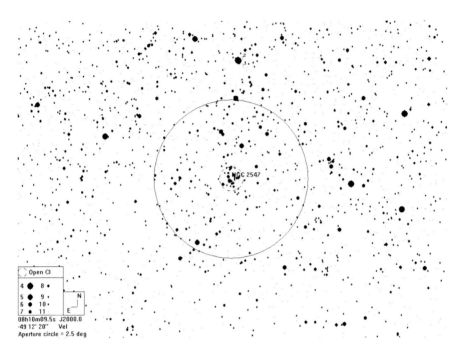

NGC 2547 is 2°S of the 2nd magnitude γ **Vel**.

NGC 2547 is a fine cluster, nearly the same apparent diameter as the Moon, that shows over 30 stars at ×37, sky conditions permitting.

Puppis: Open Cluster: NGC 2539 (100 mm)

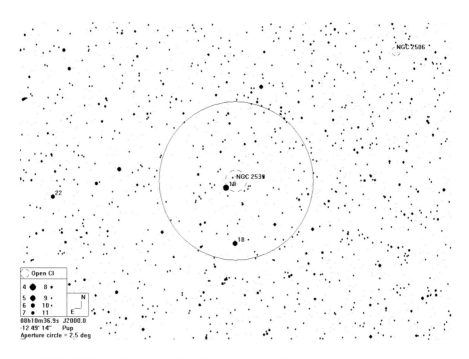

NGC 2539 is adjacent to the 5th magnitude **19 Pup**, which itself is 8° SE of α Mon.

NGC 2539 is a challenging object to locate, but relatively easy to identify. It requires good sky conditions which, owing to its declination, are rare from the latitude of Britain. I find the surrounding star field confusing for star-hopping and my usual method of location is to scan the region with 10×42 binoculars, in which it appears as a faint misty patch, and find the location with these in order to point the larger instrument in the same direction. It appears in the larger instrument merely as a larger misty patch, but one that is attractive for its delicacy.

Puppis: Open Cluster: M93 (NGC 2447) (70 mm)

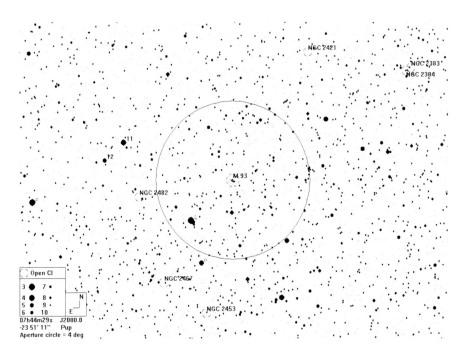

If you place ξ **Pup** at the SE of the field of view, **M93** should appear approximately in the center of the field.

M93 is a bright (magnitude 6.2), rich, and densely packed cluster in which some 25–30 stars are visible in a 15×70, with more giving a glowing backdrop. It is unusual in that the center of the cluster, which is bounded by an arrowhead-shaped grouping of brighter stars, is relatively sparse.

Puppis: Open Cluster: NGC 2451 (50 mm)

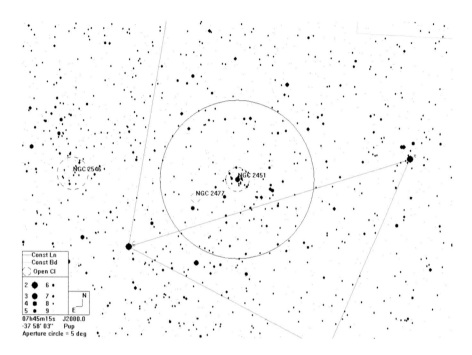

NGC 2451 is 1°N of the center of a line drawn from ζ **Puppis** to π **Pup**.

NGC 2451 is quite a sparse cluster with several relatively bright stars, which makes it a good object for binocular observation. Adding to the attractiveness of the cluster is the fact that the brightest star is orange and the surrounding bright stars are brilliant white. Six stars are particularly easy to see and a dozen or so are visible in medium binoculars under good conditions.

Puppis: Open Cluster: NGC 2477 (C71) (70 mm)

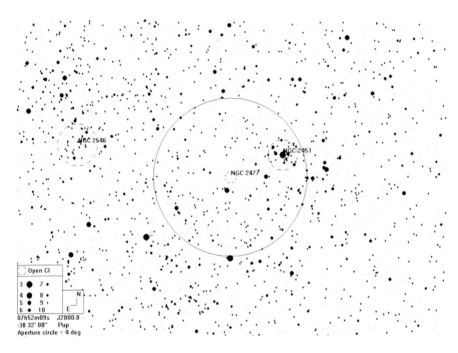

NGC 2477 is a very slightly more than one field NW of ζ **Pup**.

This is an absolutely superb cluster in a 15×70 binocular. Only a handful of stars are resolved in big binoculars, but the hundreds of unresolved stars provide a beautiful backdrop to those few. Compare it to the sparser **NGC 2451**, which can be included in the same field and is described in the list of 50-mm objects.

Puppis: Open Cluster: NGC 2546 (100 mm)

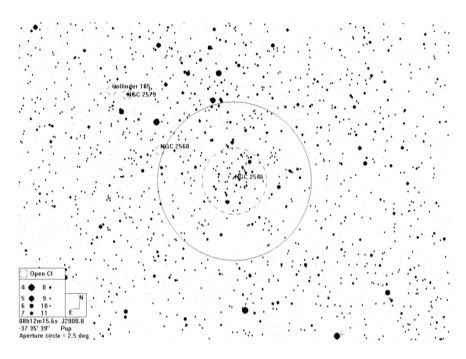

NGC 2546 is located 3° to the NE of ζ **Pup**

This huge, sparse cluster is a fine sight in big binoculars, which reveal the varied colors of some of the brighter stars. It lies in a lovely star field and is altogether a delightful object.

Hydra: Open Cluster: M48 (NGC 2548) (70 mm)

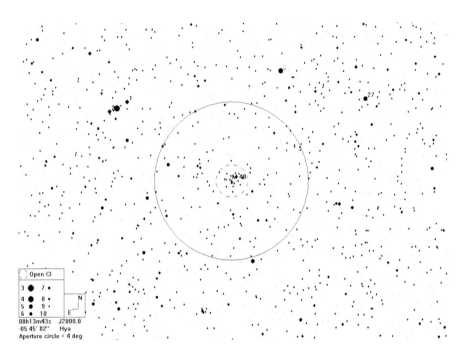

M48 is in a star-sparse region of sky and can be tricky to find until you know where it is. Firstly, identify ζ Mon (it is the easternmost, and brightest, star in an equilateral triangle of 5th-ish magnitude stars that it makes with 27 and 28 Mon). Now put ζ Mon at the NNW of the field of view, and you should be able to see the 5.8th magnitude cluster on the opposite side of the field of view, 3° to the SSE.

M48 is worth hunting down. It is brighter to the E than to the W and should show about two dozen stars in a 70-mm binocular under a good sky. Averted vision tends to bring more stars into view.

Vela: Open Cluster: IC 2391 (C85, the *Omicron Velorum Cluster*) (50 mm)

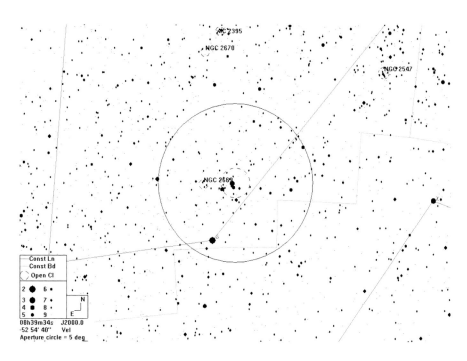

IC 2391 surrounds o **Velorum**, which is 2° NNW of δ **Vel**.

IC 2391 is a fairly sparse cluster, about twice the diameter of the Moon, in which 10 or so stars resolve in 10×50 binoculars.

Cancer: Open Cluster: M44 (NGC 2632, *Praesepe*, the *Beehive Cluster*) (50 mm)

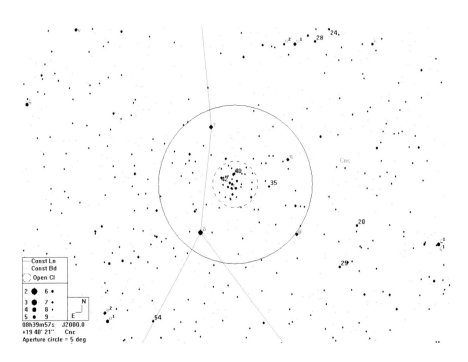

M44, which is visible to the naked eye, is in the same 5° field as γ, δ, and η **Cancri** and contains ε **Cnc**.

The Beehive is a very nice binocular object, in which you may be able to resolve up to 20 or so stars in 10×50 binoculars. You should also be able to resolve two binocular double stars, ADS 6915 and ADS 6921.

Cancer: Open Cluster: M67 (NGC 2682) (70 mm)

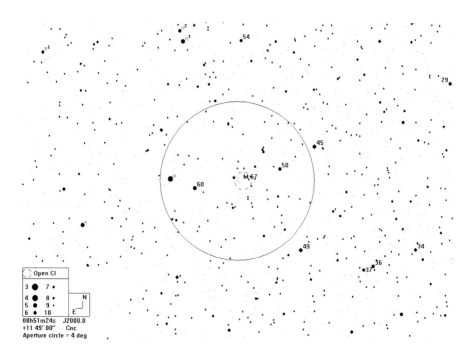

Place **Abucens** (α **Cnc**) at the E of the field of view, and **M67** will appear near the W.

Although relatively few stars are resolved in big binoculars, this is a large, bright cluster with many stars that are too faint to be seen but which contribute to the nebulous glow. It is a curiosity in that, at an estimated 4 billion years old, it is older than usual for an open cluster.

Sextans: Double Star: 9 Sextantis (100 mm)

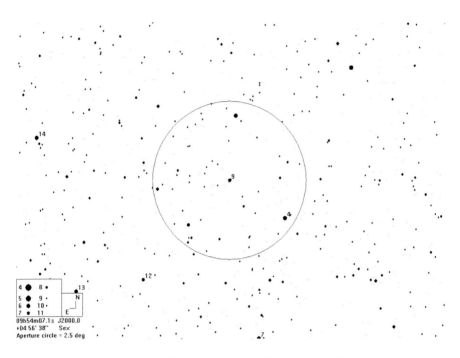

9 Sex is 8° SW of α **Leo (Regulus)**. It is in the direction of ι **Hya**, and the hop is aided by π **Leo**, which is just over halfway from α Leo to the double.

9 Sex is a widely (53 arcsec) pair of 6th and 9th magnitude stars. The primary is noticeably red.

Ursa Major: Galaxy Pair: M81 (NGC 3031) and M82 (NGC 3034) (100 mm)

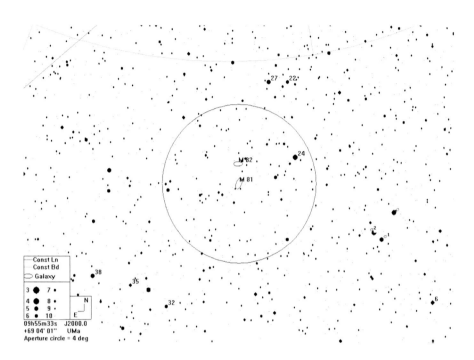

I usually find this pair with the aid of a reflex finder. Extend a line from *Phecda* (γ **UMa**) through *Dubhe* (α **UMa**) the same distance beyond *Dubhe*. From this point, the galaxies are a degree or so in the direction of *Polaris* (α **UMa**), adjacent to the 4.5th magnitude **24 UMa**.

This pair fits easily into a 2° field of view and offers a nice contrast. **M81**, also known as ***Bode's Nebula***, is the easier of the pair in small binoculars, but both are easy objects in 70-mm glasses, with **M81** clearly showing a very bright nucleus and **M82** appearing bright but mottled along its long axis, giving rise to its common name: the *Cigar Galaxy*. Their difference in orientation is obvious.

They also offer a good demonstration of averted vision: center **M81** and then look at **M82**. Notice how **M81** appears to enlarge and how its nucleus becomes more distinct. Now center **M82** and avert your gaze to **M81**: note how the mottled effect on **M82** becomes more apparent.

Chapter 12

March Equinox to June Solstice (RA 10:00 h to 16:00 h)

Carina: Open Cluster: NGC 3114 (50 mm)

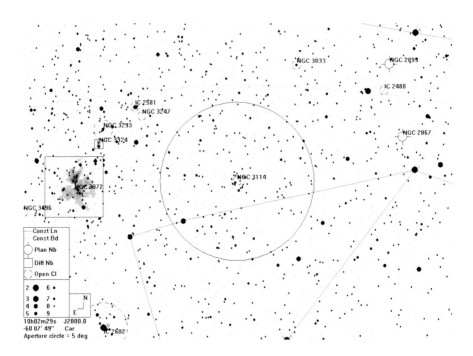

NGC 3114 is just visible to the naked eye and is situated just over 5°W of the η **Carinae** nebula (above).

NGC 3114 is a good object for 10×50 binoculars, with 15 or so stars being resolved in a region about the same size as the Moon. It is superb in larger instruments.

Sextans: Galaxy: NGC 3115 (C53, the *Spindle Galaxy*) (100 mm)

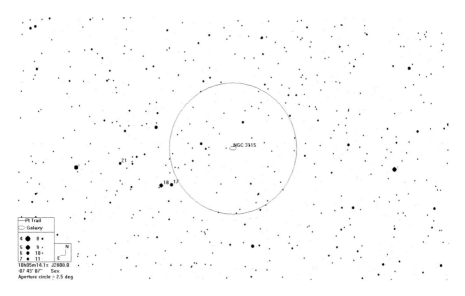

Identify the 5th magnitude γ **Sex**, which is 6° E of α **Hya** (*Alphard*). **NGC 3115** is one and a half fields further E from γ **Sex**.

100-mm binoculars at ×37 will show clearly how this bright galaxy got its name. It is extended about five times its width on a NE-SW axis.

Hydra: Planetary Nebula: NGC 3242 (C59, the *Ghost of Jupiter*) (100 mm)

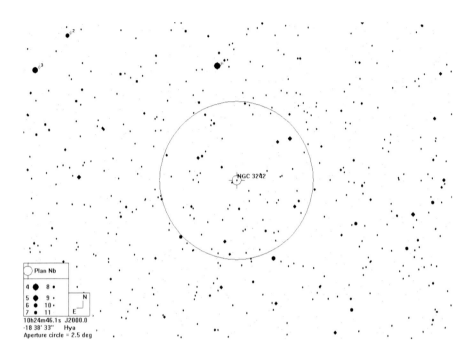

Place the 4th magnitude μ **Hya** at the N of the field and **NGC 3242** should appear about halfway to the edge on the opposite side. If you have trouble identifying it, flash a UHC or [O-III] filter across an eyepiece.

The *Ghost of Jupiter* appears stellar in smaller binoculars, but is distinctly nonstellar at 37×100, although I have not been able to detect the oblateness that gives it its common name in these.

Carina: Open Cluster: IC 2602 (C102, the θ *Carinae Cluster*, the *Southern Pleiades*) (50 mm)

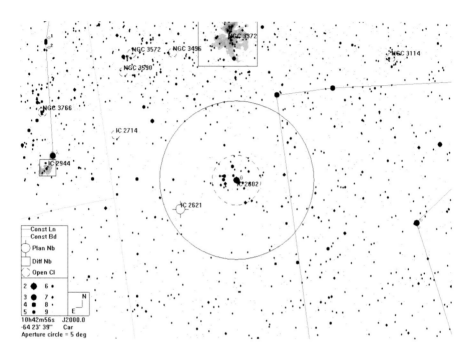

IC2602 is an easy naked-eye cluster surrounds the star θ **Carinae**.

This large cluster is similar in size and the number of bright stars as **M45**, thus giving it its common name. It is a particularly fine sight in 10×50 binoculars, with 20 or more stars, depending on sky conditions, being resolved.

Carina: Emission Nebula: NGC 3372 (C92, η *Carinae Nebula*) (50 mm)

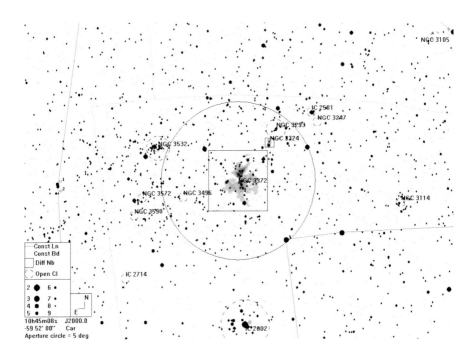

If you place **θ Carinae** at the S of a 5° field, **NGC 3372** will be at the N edge.

The *η Carinae Nebula* is one of the stunning sights of the southern hemisphere.

It is easily visible to the naked eye and begins to show detail even in small binoculars. It is a gas and dust shell that surrounds the enigmatic star η **Car**, which has fluctuated in brightness from 3rd magnitude when it was first catalogued, to nearly as bright as *Sirius* in the mid-nineteenth century (when the nebula formed), to its current status of invisible to the naked eye. The nebula and its progenitor star are 8,000 light-years away, nearly a thousand times as distant as *Sirius* and five times as far as the *Great Orion Nebula*.

Leo: Galaxy Trio: M95 (NGC 3351), M96 (NGC 3368), and M105 (NGC 3379) (100 mm)

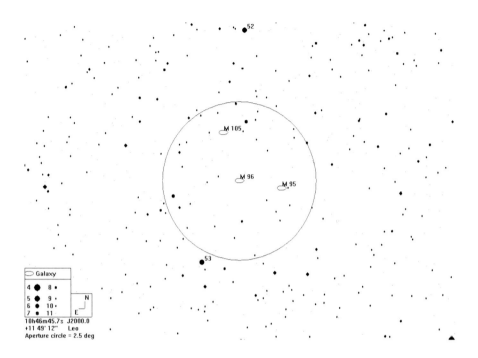

The coordinates are for **M96**.

The galaxies are to be found two fields to the NE of ρ **Leo**.

This is another galaxy trio that is neatly framed in the field of big binoculars. The two companion galaxies to **M105** (**NGC 3371** and **NGC 3373**) are tricky, but possible, bringing the total in one field to five galaxies.

Leo: Galaxy: NGC 3521 (100 mm)

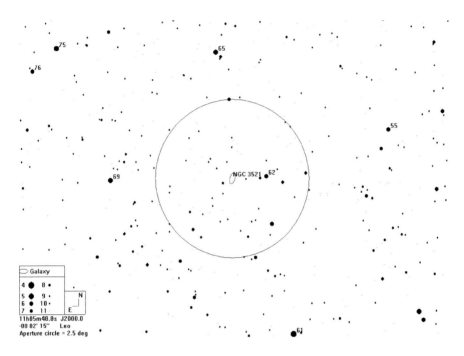

Midway between β **Vir** and β **Sex** identify the white 5th magnitude **69 Leo** and then the deep yellow **62 Leo**. **NGC 3521** is half a degree E of **62**.

This bright (10th magnitude) galaxy is often overlooked as a binocular object owing to its location away from the bright stars by which the constellation of Leo is recognized. It is considerably larger than, for example, any of the **M95/96/105** trio and is as bright as **M96**. It is clearly elongated along a SSE-NNW axis.

Leo: Galaxy: NGC 3607 (100 mm)

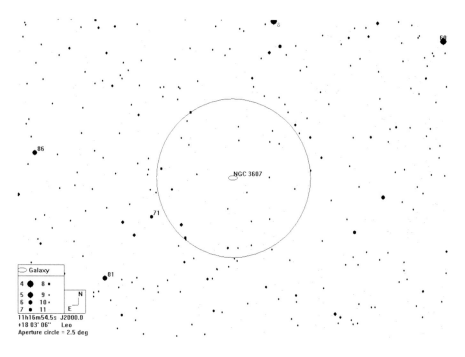

NGC 3607 is located just over 2.5° SSE of δ **Leo**.

This 11th magnitude galaxy is almost spherical. It is small and compact and, therefore, relatively easy to see. It is the central galaxy of a tight group of 3; the northern one is possible but challenging at ×37, but I have never seen the more southerly one in binoculars.

Leo: Galaxy Trio: M65 (NGC 3623), M66 (NGC 3627) and NGC 3628 (100 mm)

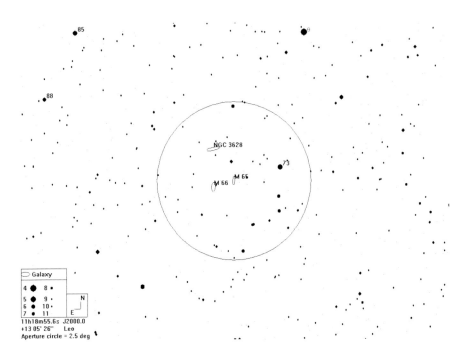

Coordinates are for **M65**.

Put θ **Leo** (*Chort*) at the N of the field and find **73 Leo** 2° to the S. Place **73 Leo** at the W of the field and the galaxies should be visible near the middle.

These galaxies are nicely framed in a 2.5° field. Although they are visible as flecks of light in 10×50 binoculars, they are distinctly better at ×37 and the difference in shape of **NGC 3628** becomes apparent.

Ursa Major: Planetary Nebula: M97 (NGC 3587, the *Owl Nebula*) (100 mm)

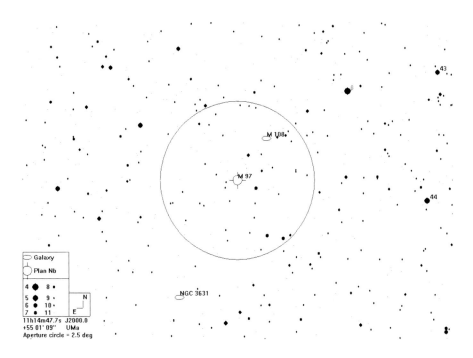

Slightly over 2° SE of **Merak (β UMa)** is a 7th magnitude star. **M97** is just under ½° ENE of this star.

This 12th magnitude planetary nebula appears as a sharp-edged disc of glowing mistiness. It is more distinct with a UHC or [O-III] filter but, even with this, do not expect to see the "eyes" that give this object its common name.

Ursa Major: Asterism: M40 (100 mm)

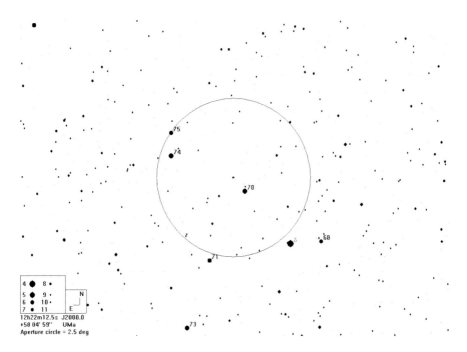

Take a line from *Phecda* (γ **UMa**) to *Megrez* (δ **UMa**) and follow it a degree further to the magnitude 5.5 star **70 UMa**. Continue another quarter of a degree and you will find **M40**.

M40 is a magnitude 9.1 optical double star whose components are separated by slightly less than an arc minute. It must surely win, by a country mile, the accolade for being the most boring Messier object: it is easy to find, easy to see, and easy to split. The mystery surrounding it is how Charles Messier thought anyone might mistake this entirely unremarkable pair for a comet!

Corvus: Planetary Nebula: NGC 4361 (100 mm)

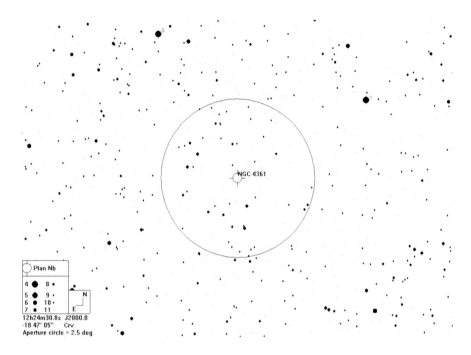

NGC 4361 is located one field diameter to the SE of γ **Crv**.

This compact, 10th magnitude planetary nebula is a difficult object from the latitude of southern Britain, owing to its low altitude of culmination, but it is considerably easier from the latitude of the Mediterranean. As with all objects of this type, it benefits from a UHC or [O-III] filter and averted vision. The 12th magnitude progenitor star is also visible at ×37.

Centaurus: Open Cluster: NGC 3766 (C97, the *Pearl Cluster*) (100 mm)

Place λ **Cen** at the S of the field and **NGC 3766** will appear near the center.

NGC 3766 is a superb cluster that is situated in a particularly rich region of the Milky Way. At ×37 it resolves into dozens of mostly blue stars, contrasted by a pair of deep red 7th magnitude stars that are like the ends of an arrow that pierces a heart-shaped asterism of fainter stars. Any of the other clusters shown on the chart are also worth visiting while in the region.

Centaurus: Open Cluster and Supernova Remnant: IC 2944 (C100, *the Running Chicken, the λ Centauri Nebula*) (100 mm)

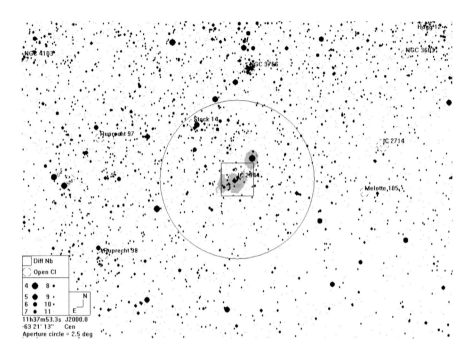

IC 2944 is located immediately to the SE of λ Cen, which some sources consider to be part of the object.

Among the few dozen stars that comprises IC 2944 is a chain of brighter blue stars running from NW to SE. Patience, coupled with averted vision, is sometimes needed to pull the nebulosity from the rich background. A UHC or [O-III] filter is an obvious aid.

Canes Venatici: Galaxy: M106 (NGC 4258) (100 mm)

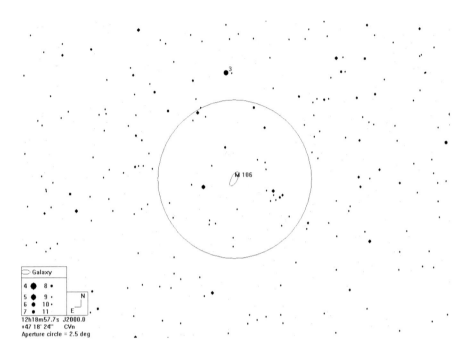

M106 is just over halfway from *Phecda* (γ **UMa**) to *Chara* (β **CVn**). Alternatively, follow the chain of 8th magnitude stars to the E of *Phecda* until you get to the obviously brighter (5th magnitude) **5 CVn**, and then head S until you get to the equally bright **3 CVn**, which is just over half a degree N of the galaxy.

This magnitude 8.3 galaxy is an easy object in binoculars of 70-mm or greater aperture. It shows elongation in an SE-NW orientation; this is noticeable (usually needing averted vision) in a 10×50 and is obvious in a 100-mm binocular, where it appears as a bright misty glow around a slightly oval nucleus.

Canes Venatici: Galaxy Pair: NGC 4631 (C32, *the Whale Galaxy*) and NGC 4656 (100 mm)

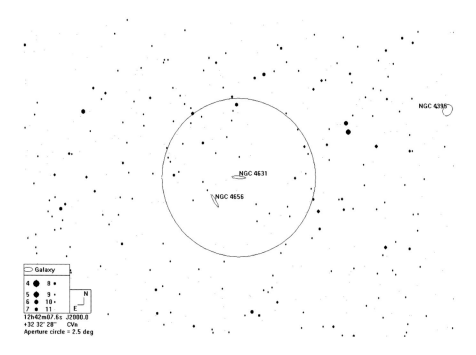

The coordinates are for **NGC 4631**.

These galaxies are almost at the midpoint of a line from **Cor Caroli** (α **CVn**) to γ **Com**, slightly towards the latter.

NGC 4631 is the easier of these two elongated galaxies, but they are both possible in 100-mm binoculars where they appear, at ×37, as a pair of short streaks of light in the sky.

Canes Venatici: Carbon Star: Y CVn (*La Superba*) (50 mm)

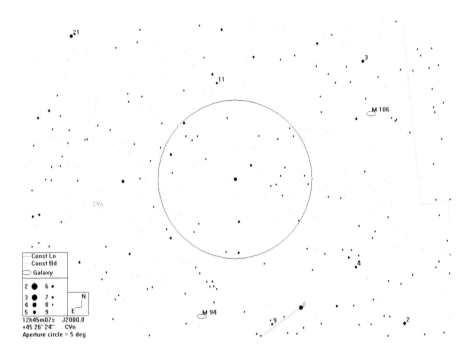

Y Canum Venaticorum is 4½° to the NW of the 4th magnitude β **Cvn**.

This 5th magnitude star was named *La Superba* by Angelo Secchi, a nineteenth-century Italian astronomer, as a consequence of its deep red color, which is brought out well by binoculars. Y Cvn is a star that is near the end of its life and shows carbon in its atmosphere. It is variable over about half a magnitude with a period of 158 days.

Canes Venatici: Galaxy: M94 (NGC 4736) (70 mm)

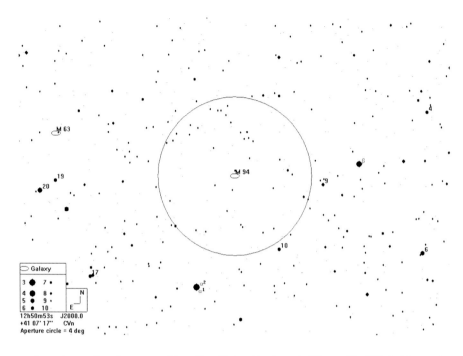

Imagine a line between ***Cor Caroli*** (**α Cvn**) and ***Chara*** (**β CVn**). From the midpoint of this line, offset two degrees (approx half a field of view) in the direction of *Alkaid* (**η Uma**), where you should find M94.

M94 is a magnitude 8.2 spiral galaxy whose light has taken 13.6 million years to reach us. It can appear like a defocused star with direct vision but will usually at least double in size with averted vision.

Canes Venatici: Galaxy: M63 (NGC 5055, the *Sunflower Galaxy*) (70 mm)

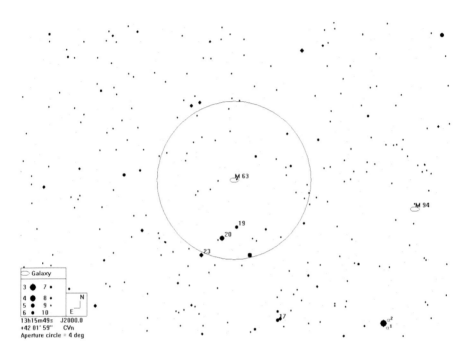

Place *Cor Caroli* (α **Cvn**) just outside the SW of the field of view of a 15×70 binocular. Near the left-hand side is a little line of three 5th magnitude stars (**19, 20,** and **23 CVn**). From **19**, go 1° towards *Mizar* (ζ **UMa**), where you will find this magnitude 8.6 galaxy.

M63, whose light has taken 37 million years to reach us, was one of the first to be identified as having a spiral structure. This is not visible in binoculars, where it appears as a short streak of mistiness extending approximately east-west.

Canes Venatici: Galaxy: M51 (NGC 5194, the *Whirlpool Galaxy*) (100 mm)

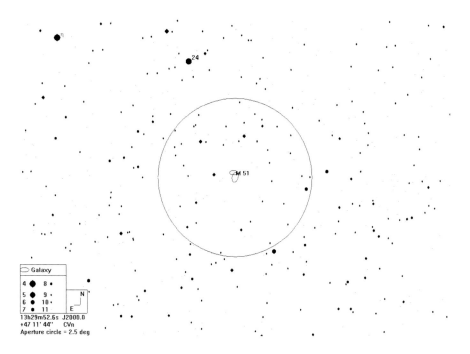

I usually find this by estimating it to be at the toe of an "L" whose upright is the line from *Mizar* (ζ **UMa**) to *Alkaid* (η **UMa**). If it is difficult to see, I hop one field WSW from *Alkaid* to **24 Cvn** and then 2° SSW. M51 is found at the NE apex of an approximate equilateral triangle that it makes with two 6th magnitude stars.

The key to seeing the **Whirlpool** is a dark transparent sky. In good conditions, it is very obvious in big binoculars, as is its companion NGC 5195 which shares the same background glow, but do not expect to see the spiral structure that was first detected by Lord Rosse in the "Parsonstown Leviathan."

Canes Venatici: Globular Cluster: M3 (NGC 5272) (70 mm)

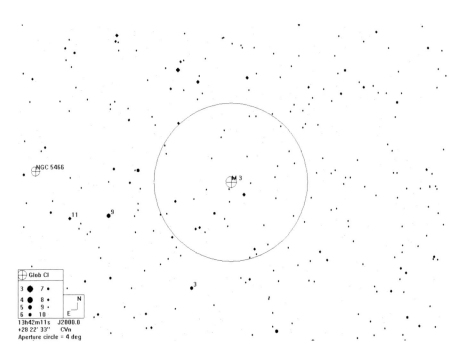

I usually find **M3** by imagining the intersection of a line from ***Cor Caroli*** (α **CVn**) to ***Arcturus*** (α **Boo**) with one from β **Com** to ρ **Boo**; if **M3** is not in the field, it is slightly towards ***Arcturus***. Alternatively, start your hop at β **Com**. M3 is 1½ fields W. It is a degree S of a line joining β **Com** to ρ **Boo**.

Although it is not large, M3 is a bright, obvious cluster which at ×15 shows distinct brightening to the core. The 6th magnitude star slightly to the SW is yellowish.

Coma Berenices: Open Cluster: Melotte 111 (50 mm)

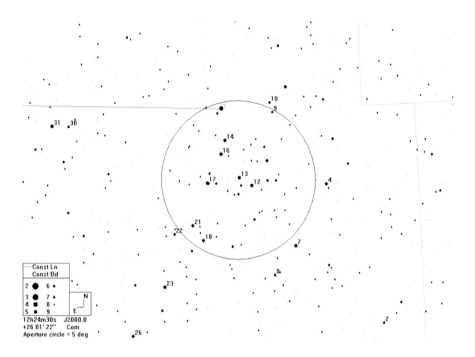

Melotte 111 is a large cluster of stars that includes γ **Comae Berenices**, although it is likely that γ **Com** is a field star, not a true member.

In a dark sky, the cluster is visible to the naked eye. It is the cluster that gives the parent constellation its name: in legend, it is the beautiful hair that Queen Berenice sacrificed to Aphrodite in order to ensure the safe return from war of her husband. This cluster, the third nearest to us, overspills a 5° field.

Coma Berenices: Galaxy: NGC 4559 (C36) (100 mm)

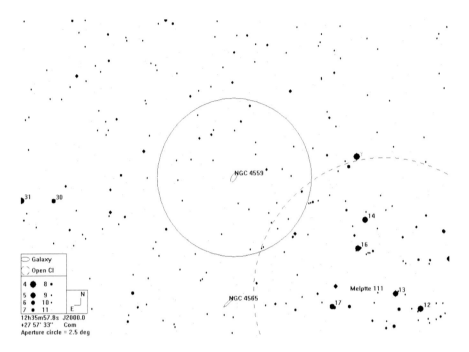

Place γ **Com** at the W of the field and **NGC 4559** will be almost diametrically opposite.

NGC 4559 is visible in smaller binoculars and is an easy object at 37×100, in which it appears only slightly oblate unless the sky is very dark, when the peripheral regions become more visible and it appears to lengthen.

Coma Berenices: Galaxy: NGC 4565 (C38, *Berenice's Hair Clip, the Needle Galaxy*) (100 mm)

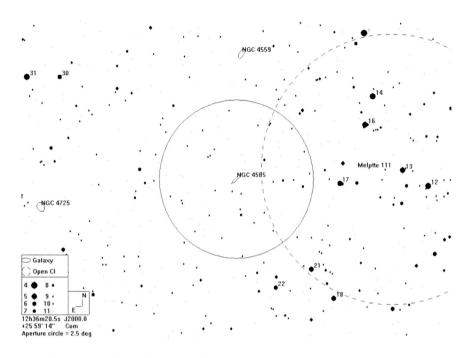

NGC 4565 is 2° S of **NGC 4559** (above)

NGC 4565 has a high surface brightness and is a lovely sight in large binoculars. Being a typical edge-on galaxy, to us it appears like a needle of light against the darker sky.

Coma Berenices: Galaxy: M64 (NGC 4826, the *Black Eye Galaxy*) (70 mm)

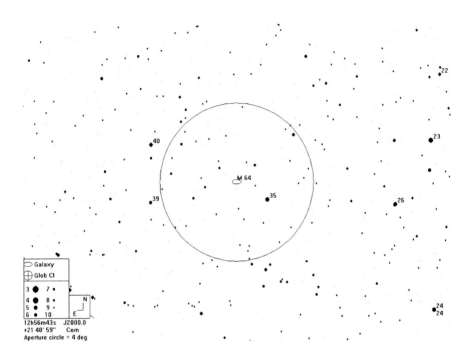

M64 lies a third of the way from **Diadem (aCom)** to **gCom**.

In 15×70 glasses, this magnitude 8.5 galaxy appears round and, although the "black eye" is not visible, it is sometimes possible to discern that the galaxy is not of a uniform brightness. This becomes more apparent with larger binoculars and higher magnification.

Coma Berenices: Globular Cluster: M53 (NGC 5024) (100 mm)

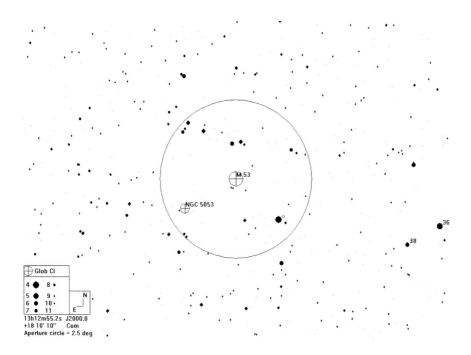

M53 lies in the same field as α **Com** (*Diadem*).

This is a relatively large globular. Although it is not as bright as the "famous" ones and requires the magnification of a telescope to resolve any stars out of it, (**M13, ω Cen**), it is quite large and is easily seen as a diaphanous glow surrounding a dense core.

Musca: Globular Cluster: NGC 4372 (C108) (100 mm)

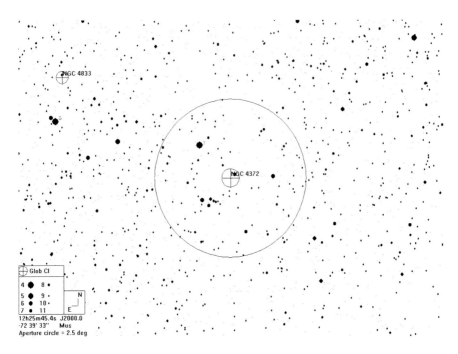

NGC 4372 is located in the same field of view as γ **Mus**, just S of the star.

NGC 4372 is large (18.6 arcmin) but of relatively faint surface brightness. Hence, it benefits from a dark, transparent sky.

Musca: Globular Cluster: NGC 4833 (C105) (100 mm)

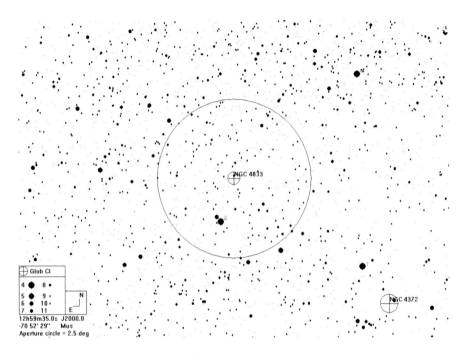

NGC 4833 is to the NNW of δ **Mus**, in the same field.

NGC 4833 is both smaller and brighter than the nearby **NGC 4372** (above). It is therefore a little easier to locate and observe, especially if sky conditions are less than ideal. It is in a rich star field on the edge of the Milky Way.

Crux: Open Cluster: NGC 4755 (C94, the *Jewel Box*) (50 mm)

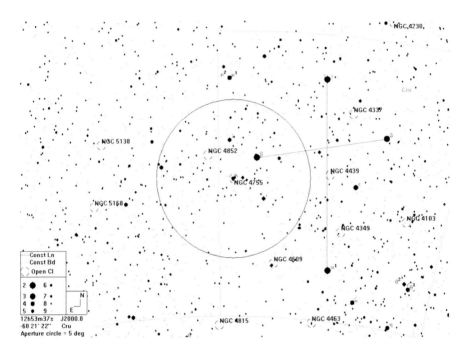

NGC 4755 is situated 1° SW of **β Crucis** and spans about 1/3 of the diameter of the Moon. It is visible to the naked eye and 10×50 binoculars will show the aptness of its common name, with several stars being resolved. With larger binoculars, you may be able to discern that one of them is reddish in color.

Virgo: Galaxy Chain: NGC 4374 (M84), 4406 (M86), 4438, 4473, 4477, and 4459 (*Markarian's Chain*) (100 mm)

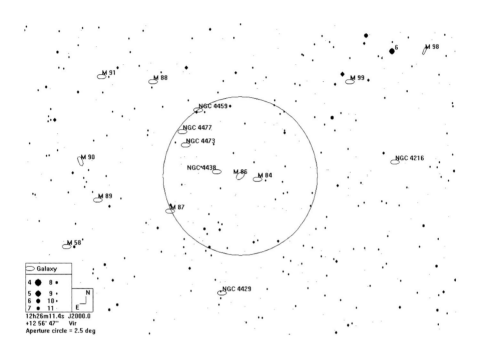

The coordinates are for M86, the brightest galaxy in the chain.

Markarian's Chain lies almost exactly halfway between β **Leo** and ε **Vir**; a reflex finder is ideal for placing you in the correct location.

The problem in this region of the sky is not finding galaxies but in sorting them out. There are tens of galaxies available to big binoculars in this region, the *Virgo/Coma* cluster. *Markarian's Chain* is a string of a dozen or so galaxies that extends over nearly 2° from Virgo into Coma. The brightest members are **M84** and **M86**, both of which are easy objects which show brightening to the core with averted vision. The chart shows the six brightest members of this chain, and you should be able to see all of these if the sky is reasonably dark. Fainter members will become visible under ideal conditions. It is worth exercising patience (and averted vision) to tease the fainter members into visibility.

While you are in the locality, it is worth panning around and seeing what other galaxies you can see—and identify!

Virgo: Galaxy: M49 (NGC 4472) (70 mm)

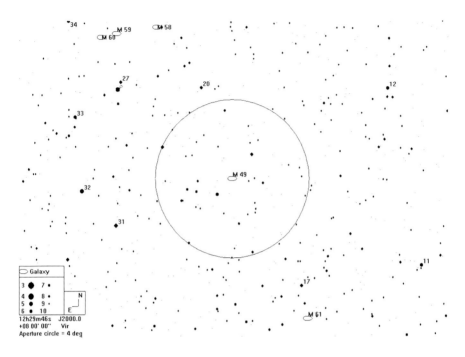

Place ρ **Vir** on the NE of the field of view; on the opposite side of the field is a pair of 6th magnitude stars separated by just over a degree. **M84** is slightly SE of the midpoint of these two stars.

This magnitude 8.4 galaxy appears as a circular patch of light that could easily be mistaken for a globular cluster. In good conditions, averted vision may show some brightening of an almost stellar-looking nucleus.

Virgo: Galaxy Group: M87 (NGC 4486) and Friends (70 mm)

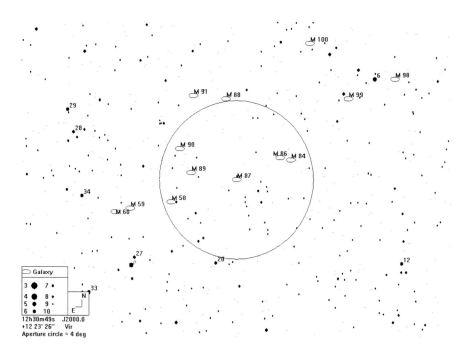

Running W from **ρ Vir** is a 2° chain of five 8th and 9th magnitude stars. The same distance further on is a pair of 8th magnitude stars orientated N-S. Half a degree NE of the northernmost of this pair of stars is a 9th magnitude star that is immediately N of **M87**.

Magnitude 8.6 **M87** is the easiest of eight Messier galaxies that can be held in the same field of view. With averted vision it will show brightening towards the center, a slight E-W extension, and a stellar-looking core.

In order of ease of observation, the other galaxies in the field are:

M86 (mag 9.2) and **M84** (mag 9.3) are part of *Markarian's Chain* and are described under that entry.

M89 (mag 8.9) may initially require averted vision, but seems to become more visible the longer you look at it.

M90 (mag 9.5) will almost certainly require averted vision, which will reveal some N-S elongation.

M88 (mag 9.5) needs averted vision or larger binoculars and shows some SE-NW lengthening.

M91 (mag 9.8) also needs averted vision with 15×70—it really needs 100 mm—but, despite its magnitude, is no more difficult than M88.

M58 (mag 10.2). I find this very difficult in 15×70, but it is possible with averted vision on a transparent night. It is easier in 100-mm glasses.

(M88 and 91 are actually in Coma)

Virgo: Galaxy Pair: M59 (NGC 4621) and M60 (NGC 4649) (70 mm)

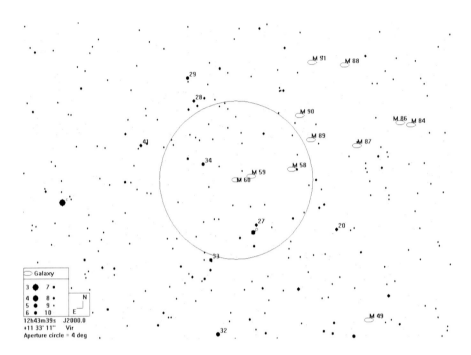

This pair of galaxies is just under a degree and a half N of ρ **Vir**.

The magnitude 8.8 **M60** is easy, but **M59** is a magnitude fainter and may require averted vision if your skies are not dark and transparent. Both could easily be mistaken for globular clusters (as is the case with many elliptical galaxies with this size of binocular), and neither shows any brightening to the center, even with averted vision.

Virgo: Galaxy: M104 (NGC 4594, the *Sombrero Galaxy*) (100 mm)

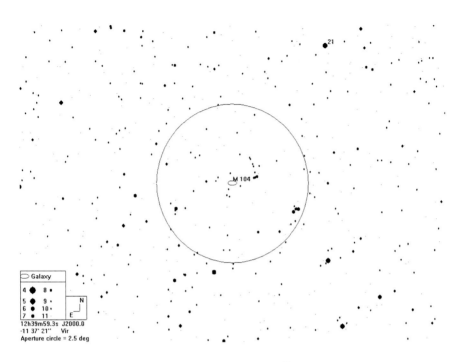

M104 is located a little over two fields NNE of δ **Crv**.

The ***Sombrero*** is distinctly elliptical with a central bulge and brighter nucleus at ×37. In ideal conditions and with averted vision, the dust lane that appears in photographs is suspected.

Hydra: M68 (NGC 4590) (100 mm)

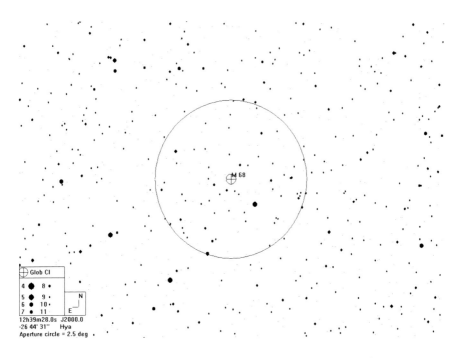

This magnitude 7.3 globular cluster lies 3½° to the SE of **Kraz (β Crv)**.

M68 is possible in 70 mm and easy in 100 mm. Like most globular clusters, it appears to grow as you switch from direct to averted vision.

Hydra: Galaxy: M83 (NGC 5263) (100 mm)

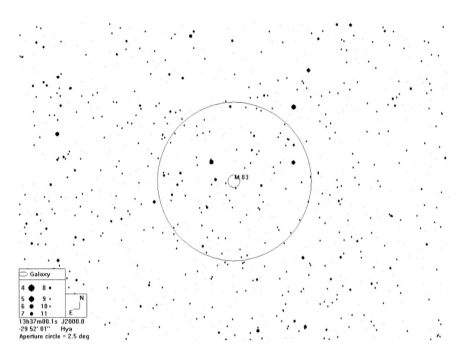

This galaxy is situated in a star-sparse region of sky. A relatively simple star-hop is available to observers whose location is sufficiently far south for **θ Cen** to be available. Starting at **θ Cen**, go 1½ fields NW to **2 Cen**. Place **2 Cen** at the SSE of the field and locate 1 **Cen** about ¾ of the way across the field. Continue this line for a further 1½ fields and M83 should be visible.

For those of a more northerly location such as most of Britain, **M83** will be a difficult object owing to its low transit altitude. Start at **γ Hya** and go just over half a field SSE to a 7th magnitude star. A whole field SE of this is another star of similar brightness. A further field to the SSE brings us via yet another star of similar brightness to a 6th magnitude star. Continue this line for a further half a field between two more 6th magnitude stars and **M83** should be visible just E of center.

M83 is a relatively bright galaxy that is almost face on. In binoculars it appears as a circular glow. It is of interest in that it has been a somewhat prolific source of supernovae, hosting no fewer than 6 during the twentieth century, several of which have been within the range of big binoculars.

Centaurus: Galaxy: NGC 5128 (C77, *Centaurus A*) (100 mm)

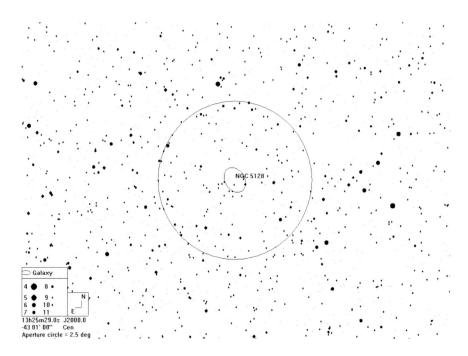

Sweep 2 fields due N of the ω **Cen** cluster (until you are due W of the more southerly of the 3rd magnitude pair μ and ν **Cen**) and this galaxy should appear almost center field.

NGC 5128 is one of the better binocular galaxies, being bright and extended. It seems to elongate more with averted vision, when the dark lane that crosses it also becomes visible. It is the location of the radio source *Centaurus A* and is also a strong X-ray source. It is thought to be two galaxies in collision.

Centaurus: Globular Cluster: NGC 5139 (C80, Omega Centauri) (50 mm)

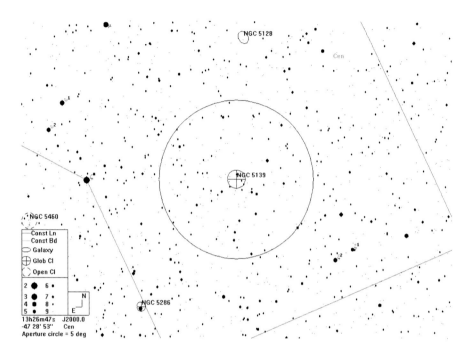

This cluster is visible to the naked eye one 5° field to the E of ζ **Centauri**.

Called the "King of Globulars" for a good reason, this globular cluster is superb in any instrument. It is noticeably larger than the Moon and extremely bright. It contains about a million stars.

Ursa Major: Galaxy: M101 (NGC 5457) (100 mm)

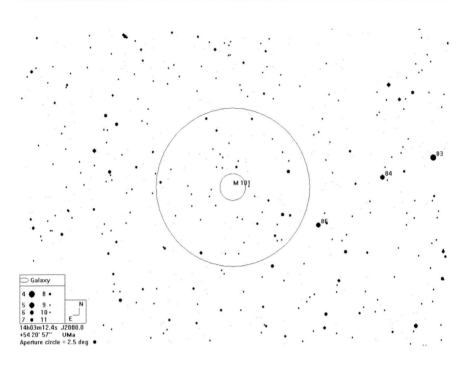

M101 is just inside the 3rd apex of an equilateral triangle that has *Alkaid* (η UMa) and *Mizar* (ζ UMa) as its other apexes.

M101 has a visual integrated magnitude of 7.7, but is very large in apparent area (nearly the size of the Moon) and, consequently, a low surface brightness. It requires a dark, transparent sky. It appears as a circular brightening of the sky. It is quite easy once you know what you are looking for, but can be tricky the first time. Use averted vision and tap the binocular to make it jiggle slightly, and you should see it.

Draco: Galaxy: NGC 5866 (100 mm)

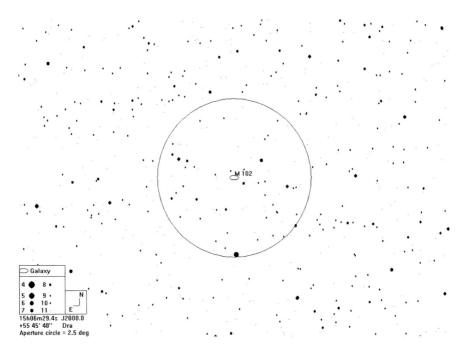

NGC 5866 is 4° SSW of ι **Dra**.

The magnitude 9.9 **NGC 5866** is the galaxy that is the prime candidate for the disputed **M102** (the other possibility is **M101**). The disputes do not end there: **NGC 5866** is a lenticular galaxy (Type SO_3), but is frequently classified as an elliptical. At ×37 it shows a very slight elongation. Also look for **NGC 5907** a degree and a half to the NE.

Draco: Galaxy: NGC 5907 (the *Splinter Galaxy*) (100 mm)

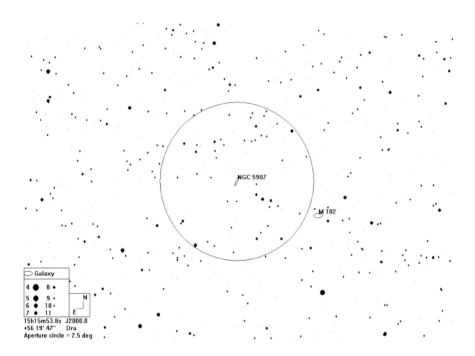

NGC 5907 lies just under a degree and a half NE of **NGC 5866**.
This magnitude 10.7 edge-on galaxy benefits from a good sky, so wait until it and the transparency are high, when it will become apparent as a thin wisp of light extending from SE to NW.

Boötes: Variable Star: RV Boötis (100 mm)

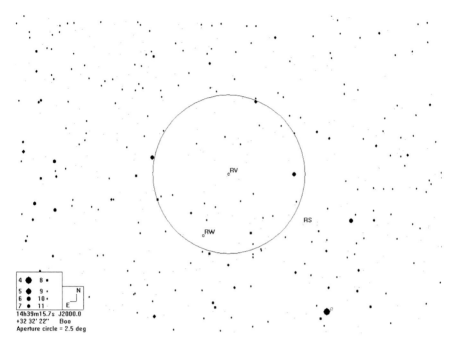

RV Boo is just over one field to the NE of **ρ Boo**:
Mean range: 7.9–9.8
Mean Period: 137 days
Type: Semi-regular
Also in the field is **RW Boo**:
Mean Range: 8.0–9.5
Mean Period: 204 days
Type: RRCrB

Boötes: Multiple Stars: δ Boötis and 50 Boötis (100 mm)

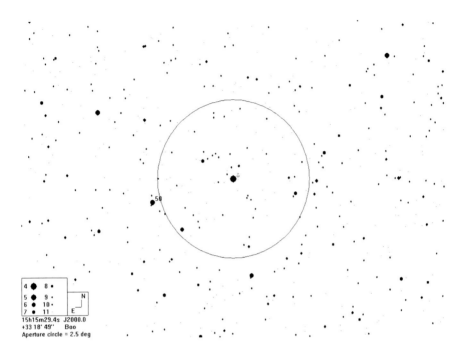

The coordinates are for **δ Boo**.

δ Boo is a double star. The yellow-white 8th magnitude secondary is 105 arcsec from the deeper yellow **δ Boo**. Large binoculars show the color difference more distinctly.

50 Boo is a triple star. The primary is a 5th magnitude blue-white, and the 10th magnitude members are a noticeable yellow, making a pretty trio.

Serpens: Globular Cluster: M5 (NGC 5904) (70 mm)

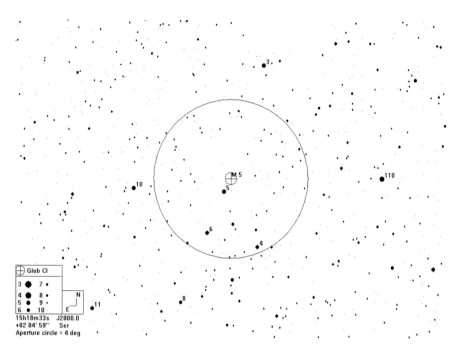

Find **M5** by panning just under two fields from α **Ser** in the direction of μ **Vir**. **M5** is one of the better globulars for binoculars. Although it does not seem as bright as **M13**, it appears slightly larger.

Chapter 13

June Solstice to September Equinox (RA 16:00 h to 22:00 h)

Triangulum Australe: Open Cluster: NGC 2065 (100 mm)

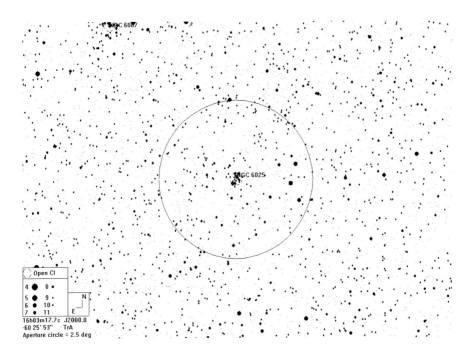

Follow the line that joins γ to ε to β **TrA** a further 3° and **NGC 2065** will be easily visible.

This bright cluster resolves into a dozen or more stars in big binoculars. The brighter members are a distinct brilliant diamond-white against a delicate glow.

Norma: Open Cluster: NGC 6067 (100 mm)

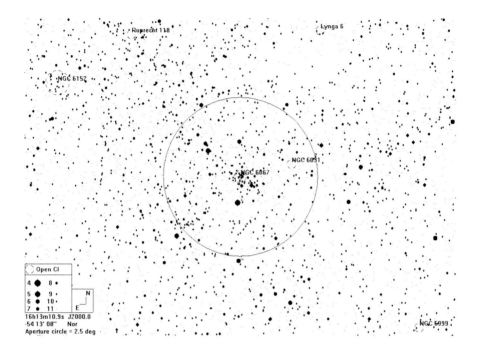

NGC 6067 is a Moon-diameter N of κ **Nor**.

Situated in a rich part of the Milky Way, this is a very impressive cluster in big binoculars, showing distinctly different colors among its brighter stars.

Scorpius: Globular Clusters: M4 (NGC 6121) and NGC 6144 (70 mm)

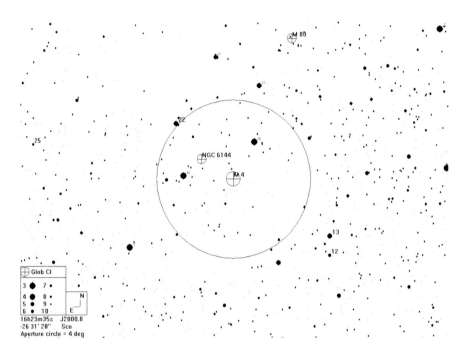

M4 has to be the easiest globular cluster to find. Simply place Antares (α **Sco**) on the E of the field and **M4** will lie at the center.

M4 is easy to identify in any binoculars. It is relatively close to us (7,000 light-years) for a globular cluster; indeed, it is closer than some open clusters. Owing to its proximity, it appears as a rather loose cluster that begins to reveal detail, including a brighter central bar, in 15×70 glasses. It would be even more spectacular were it not for intervening dust. It is in a beautiful star field that I find easier to appreciate in binoculars than in a telescope.

In the same field is the fainter and smaller **NGC 6144**, which requires good conditions to be recognized as a globular cluster in this size binocular.

Scorpius: Open Cluster: NGC 6231 (C76) (50 mm)

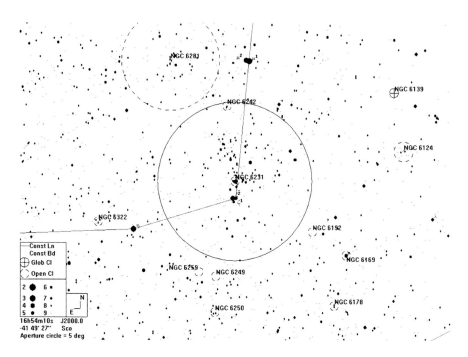

NGC 6231 is visible to the naked eye and lies half a degree north of ζ **Scorpii**. This often-overlooked cluster is a fine object in 10×50 binoculars. It is small, with a diameter about half that of the Moon, but rich in brighter stars, several of which are resolved in small and medium binoculars.

Scorpius: Open Cluster: NGC 6322 (100 mm)

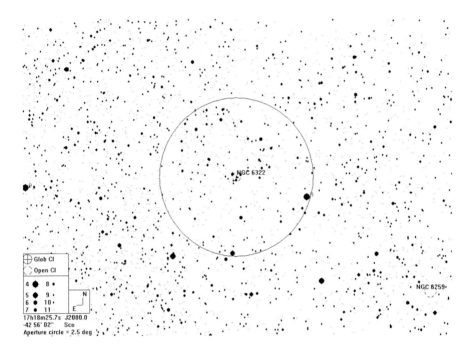

Place η **Sco** near the W edge of the field and **NGC 6322** will appear near the center.

The stars of this pretty cluster are framed by a near-equilateral triangle of stars of about magnitude 7.5.

Scorpius: Open Cluster: M6 (NGC 6405, the *Butterfly Cluster*) (50 mm)

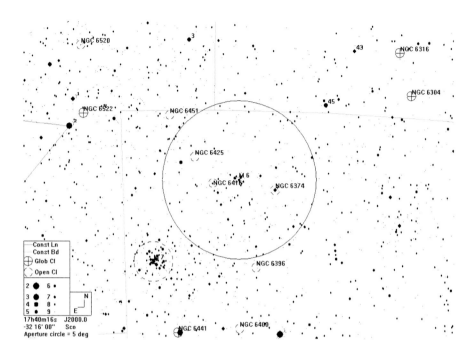

M6 is easily visible to the naked eye one 5° field N of λ **Scorpii**.

While not as impressive as its neighbor, **M7**, **M6** is a fine object in 10×50 binoculars, with half a dozen or so of the brighter stars being resolved. I find that slightly more magnification is necessary to enable the "butterfly" shape, from which it acquires its common name, to become apparent.

Scorpius: Open Cluster: M7 (NGC 6475, *Ptolemy's Cluster*) (50 mm)

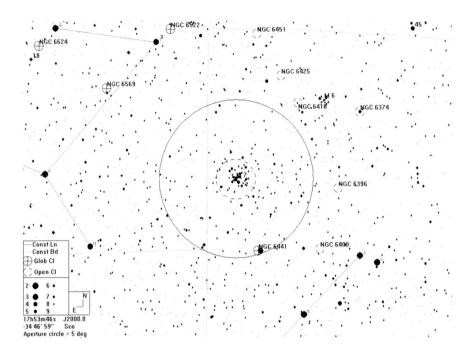

This cluster, which is visible to the naked eye as a fuzzy patch against the background Milky Way, is just over 4½° from **λ Sco** (*Shaula*), the scorpion's sting; to the medieval Arabs it was known as the Scorpion's venom.

M7 is a very large, bright cluster, about 2½ times the diameter of the Moon, in which binoculars of any size will reveal individual stars, about 9 of which are visible in 10×50 binoculars from a reasonably dark site, up to a dozen if the site is very dark. Greater magnification reveals more stars, about 80 of which are 10th magnitude or brighter. This fine cluster derives its common name from the observation of it by Ptolemy of Alexandria in the first century A.D.

Ophiuchus: Triple Star: ρ Ophiuchi (100 mm)

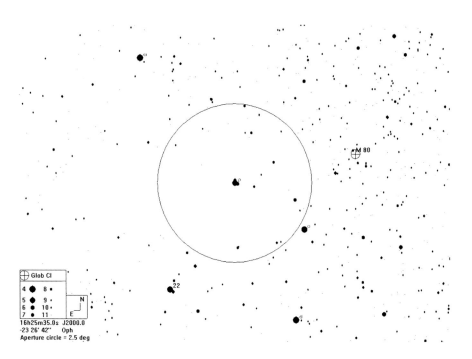

ρ Oph is one field to the NNE of **σ Sco** (the direction of **ω Oph**).

This 5th magnitude star is one of a visual triple star, whose 7th magnitude comites are situated 2.5 arcmin to the N and W, respectively. Can you see the slightly bluer color of the W comes?

Ophiuchus: M12 (NGC 6218) (70 mm)

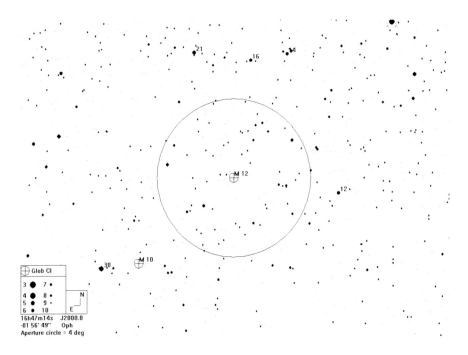

M12 is very close to the NE apex of an equilateral triangle that has δ and ζ **Oph** as its other apexes.

At magnitude 6.6, **M12** is an easy object in 70-mm binoculars in moderately good skies. Its core is less distinct than that of the nearby **M10**, with which it should be compared, not least because M12 is useful marker for finding M10, which is just over 3° to the SE. Like all globulars, it benefits from averted vision.

Ophiuchus: M10 (NGC 6254) (70 mm)

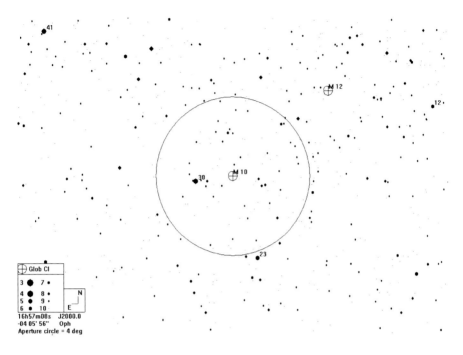

I find the easiest way to find **M10** is to navigate from **M12**. Place **M12** at the NW of the field of view, and **M10** should be near the opposite side of the field.

At magnitude 6.6, **M10** is an easy object in 70-mm binoculars in moderately good skies. It shows a very distinct brightening to the core as compared with the nearby **M12**. The bright star to the E is the 5th magnitude **30 Oph**; if your sky is sufficiently dark, you can use this star as a guide to the location of the cluster. Like all globulars, it benefits from averted vision.

Ophiuchus: M62 (NGC 6266) (100 mm)

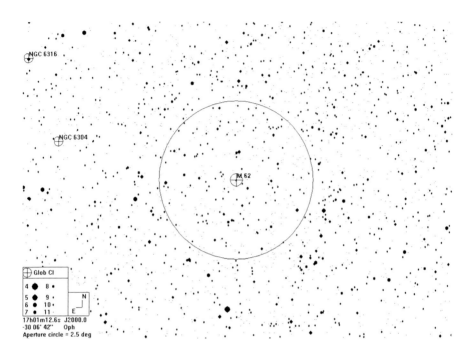

M62 lies just west of a line from θ **Oph** to ε **Sco**, about two thirds of the way along the line.

M62 is a bright (magnitude 6.4) globular that lies in a very rich star field. Like all globulars, it will seem to grow as you move from direct to averted vision, when the core will appear brighter. There are some dark lanes in the surrounding Milky Way, especially just to the E of the cluster.

Ophiuchus: M19 (NGC 6273) (70 mm)

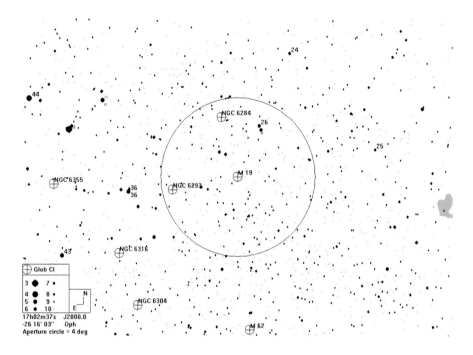

M19 is N of **M62** and ESE of **θ Oph**, with which two objects it makes the third apex of a (nearly) isosceles triangle. It can be held in the same field as **M62**.

M19 is a very bright cluster (magnitude 6.8) that lies in a very rich star field. Averted vision has an unusual effect: it changes it from round to oblate, as it seems to stretch slightly in a N-S direction. This effect is more pronounced in 100-mm binoculars but is still present in 70-mm under good skies. Its core also seems proportionally larger than is the case in the other Ophiuchus clusters, which reduces the amount of "halo growth" that averted vision usually brings.

13 June Solstice to September Equinox (RA 16:00 h to 22:00 h)

Ophiuchus: M14 (NGC 6402) (70 mm)

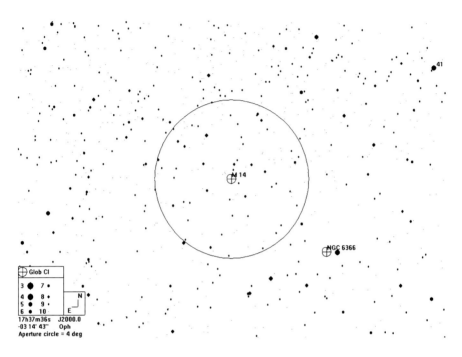

M14 lies approximately 6° (one and a half fields) W of ζ **Ser**.

This magnitude 7.6 cluster is a moderately easy object in 70-mm binoculars. Like all globulars, it benefits from averted vision, where it will appear to grow, show the brightening of the core, and possibly have a slightly triangular appearance. This latter effect, if you see it, is an optical illusion that is caused by faint strings of stars that seem to run from the SE and SW periphery of the cluster.

Ophiuchus: Open Cluster: IC 4665 (*the Summer Beehive*) (70 mm)

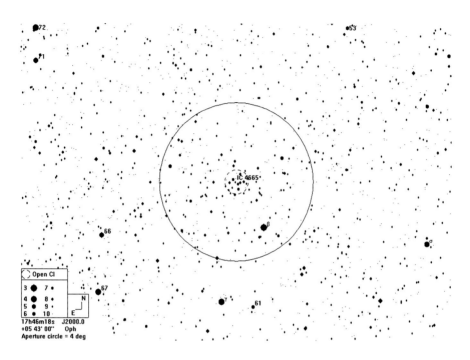

IC 4665 is half a field to the NNE of β **Oph**.

This large cluster is another object that is frequently given in lists for smaller binoculars but which benefits tremendously from larger apertures and the higher magnification that permits more stars to be revealed. Particularly attractive is a curved chain of bright white stars. This chain is part of the star-party appeal of this cluster: it forms part of the letter "H" of the word "HI."

Ophiuchus: Star: *Barnard's Star* (70 mm)

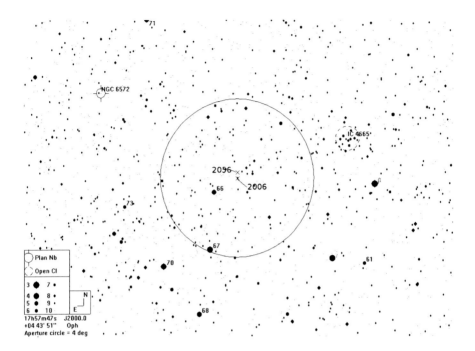

Barnard's Star is the star with the greatest known proper motion (10.28 arcsec/year). The chart gives its position for January 1 in 2006 and 2056, respectively. The large proper motion was discovered by E. E. Barnard in 1916. If you observe this 9.5th magnitude star in company, be sure to take the opportunity to dispel the notion that it has planets. This notion was the result of some shortcuts in data reduction taken by Peter van der Kamp in the 1960s and 1970s, and, although modern methods have shown it to be in error, it has gained some renewed currency on the Internet.

Ophiuchus: Open Cluster: Melotte 186 (50 mm)

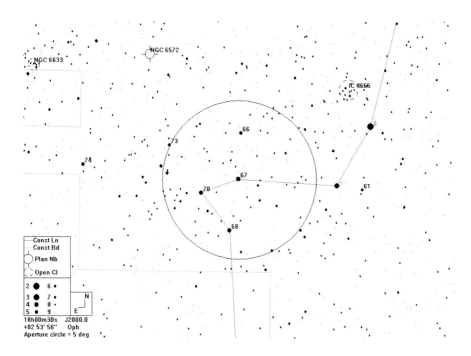

The region around **67 Ophiuchi** is a sparse open cluster, which is about 4° in diameter and includes **66**, **68**, and **70 Oph**. Look also for the smaller and denser open cluster, **IC 4665**, 4½° to the NW.

Ophiuchus: Planetary Nebula: NGC 6572 (100 mm)

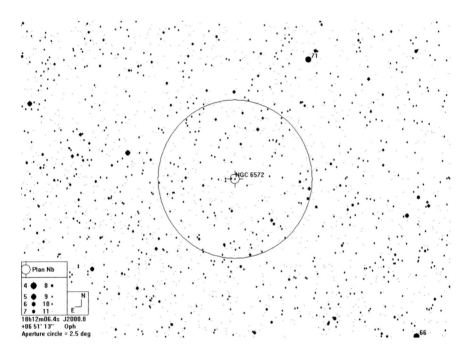

The easiest hop to this object begins at β **Oph**. Pan three fields to the E and then one field N, where you should be able to identify the star field in the chart above.

This tiny planetary appears stellar in nature but is distinguished by the beautiful green color that warrants its inclusion in this list. It is probably the greenest object that is visible in binoculars of this size.

Ophiuchus: Open Cluster: NGC 6633 (100 mm)

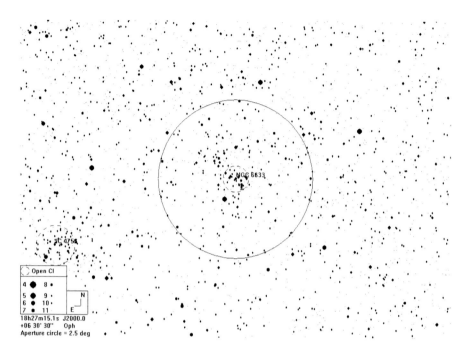

This superb cluster is often overlooked and excluded from binocular lists because it can be tricky to find. Just over 8° ESE from *Rasalhague* (α **Oph**) is the 4th magnitude **72 Oph** with the slightly dimmer **71 Oph** a degree to the S of it. From **71 Oph**, carefully pan one and a half fields to the SE and find a yellowish star that is about half the brightness of **71 Oph**. Place this star on the NW periphery of the field and the cluster should be visible towards the SE edge.

The stars in this cluster are older, and therefore more yellow, than those in many open clusters. Over 20 of these are easily resolved at ×37.

Hercules: Globular Cluster: M13 (NGC 6205) (50 mm)

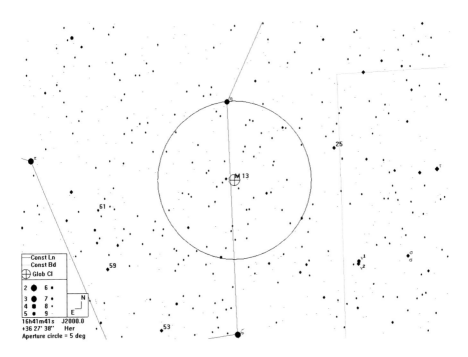

This is possibly the easiest globular cluster to find, being very bright and lying 2/3 of the way from η to ζ **Herculis**. If you place η **Her** at the N of a 5° field, M13 will lie at the center.

Although 10×50 binoculars will not resolve any stars of this cluster, its bright glow should span about 20 arcmin and may be visible to the naked eye, which is how it was spotted by Edmund Halley, who was the first to record it. Using large telescopes, several tens of thousands of stars have been resolved around the periphery of the 145 light-year wide cluster, but the stars at the core are too close together, separated by 0.1 light-years or so, to be resolved.

Hercules: Globular Cluster: M92 (NGC 6341) (100 mm)

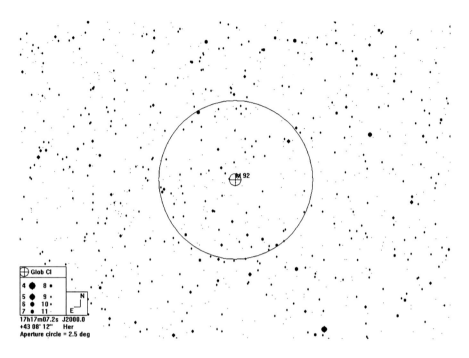

M92 lies 2/3 of the way from η to ι **Her**.

M 92 is an object that would be far better known, and more often observed, were it not for its famous neighbor, **M13**. It is a superb globular cluster in its own right, very bright with edges beginning to resolve at ×37.

Ara: Globular Cluster: NGC 6397 (C86) (100 mm)

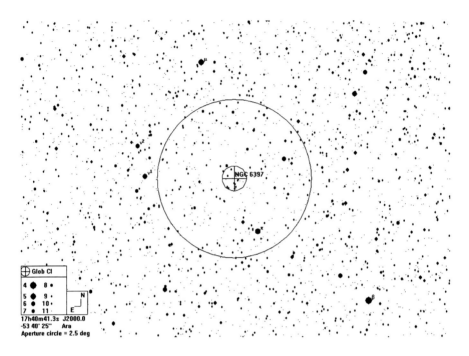

One field NE of β **Ara** is the 5th magnitude π **Ara**. Place P near the S of the field and the globular should be just off-center.

The periphery of this large, bright globular cluster begins to resolve at ×37. It is one of the nearest globular clusters.

Corona Australis: Globular Clusters: NGC 6541 (C78) and NGC 6496 (100 mm)

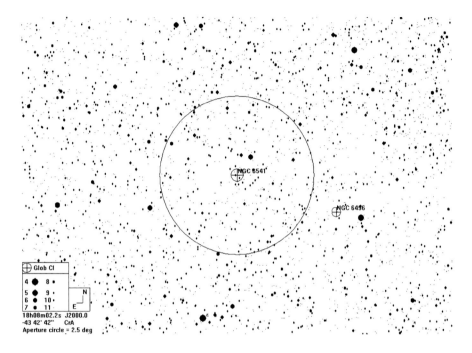

NGC 6541 is about a quarter of a field S of the point where a line from **θ CrA** to **θ Sco** crosses a line from **α Tel** to **ι¹ Sco**.

NGC 6541 is a large, bright object that is easy in all binoculars. More of a challenge is the nearby **NGC 6496**, nearly 2° to the WSW, which is about half the size and a quarter of the brightness. The 5th magnitude star nearly half a degree to the WSW of the fainter cluster can be an aid to locating it when sky conditions are less than ideal.

Sagittarius: Open Cluster: M23 (NGC 6494) (70 mm)

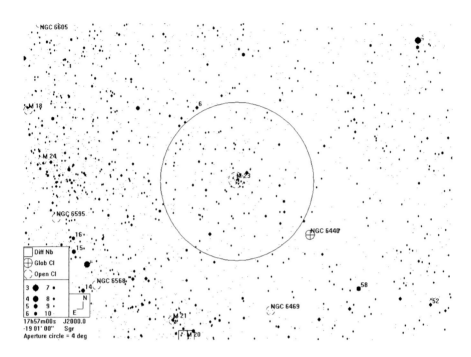

Scan SE two fields from ξ **Ser**, in the direction of μ **Sgr**. **M23** should lie near the middle of the second field.

This large bright cluster is an exquisite object in large binoculars and shows around a dozen stars at 15×70. I fancy that the brighter stars form a lower case alpha (α).

Sagittarius: Emission Nebula: M20 (NGC 6514, the *Trifid Nebula*) (100 mm)

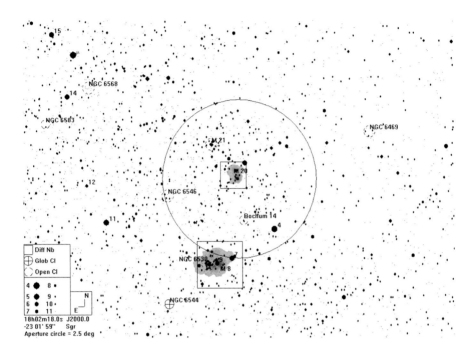

First, find the **Lagoon Nebula (M8, NGC 6523)**, which is just over 5° (2 fields) WNW of λ **Sgr**, the "peak" of the lid of the Sagittarius "teapot" asterism. Place the **Lagoon** at the S of the field and the **Trifid** should appear near the center.

In size, the **Trifid** (so-called, not because of any relation to John Wyndham's sentient plants, but because of its division into three parts by dark dust lanes) is dwarfed by **M8** to the S, but it is otherwise an impressive object. Big binoculars will resolve a handful of stars against the bright nebulosity. If it is high up in a dark sky, you may detect a greenish tinge to the nebulosity.

Sagittarius: Open Cluster and Nebulosity: NGC 6530 and M8 (NGC 6523, the *Lagoon Nebula*) (50 mm)

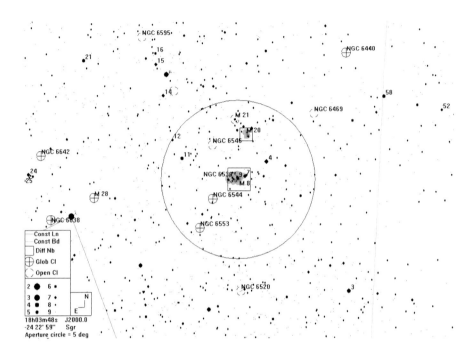

M8 is just over 5° WNW of λ **Sagittarii**, the "peak" of the lid of the Sagittarius "teapot" asterism.

The *Lagoon Nebula* is visible to the naked eye if the sky is reasonably dark and transparent. Even small "compact" binoculars, this stunning object will show a few stars of the associated open cluster (**NGC 6530**), and 10×50s will show more than half a dozen stars and some of the surrounding nebulosity (**NGC 6523**) that they illuminate, as well as the denser cluster of stars to the E of the main nebulosity. The nebulosity benefits greatly from averted vision. To the N, and encompassed by the same 5° field of view, is the smaller and fainter **M20 (NGC6514)**, the *Trifid Nebula*. This entire region of the sky is worth scanning for other "fuzzy blobs," of which there are many that are visible in binoculars of all sizes.

Sagittarius: Star Cloud: M24 (50 mm)

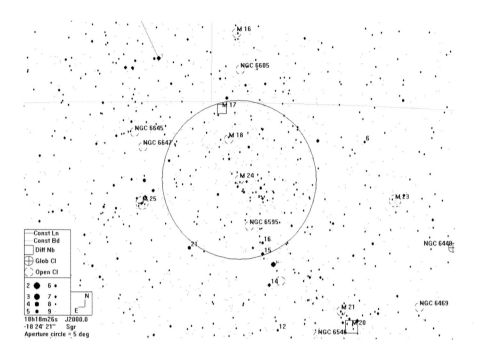

The Sagittarius Star Cloud, **M24**, lies slightly more than halfway from **γ Sgr** to **μ Sgr**.

M24 is a bright patch of light that is easily visible to the naked eye and, from southern England, has even been mistaken for a cloud on the horizon! Even small compact binoculars begin to reveal detail and it is a remarkably good object in 10×50s. Look for the open cluster **NGC 6603** (an object to which the designation M24 is often falsely ascribed) as a brighter patch in the NE of the cloud.

Sagittarius: Open Cluster: M18 (NGC 6613) (100 mm)

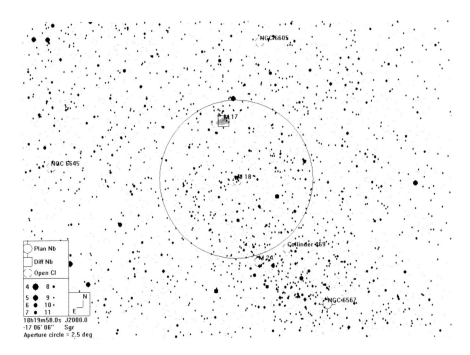

M18 is a degree S of **M17** (below).

About 15 stars are visible in this rather sparse, magnitude 6.9 open cluster, but it is worth spending time in its environment, which is pregnant with small dark nebulae. By comparison, one of the other open clusters shown in the chart, **NGC 6605**, is far less impressive: it is entirely nonexistent! It is one of several objects in the NGC catalogue that does not actually exist.

Sagittarius: Emission Nebula: M17 (NGC 6618, the Omega Nebula or Swan Nebula) (100 mm)

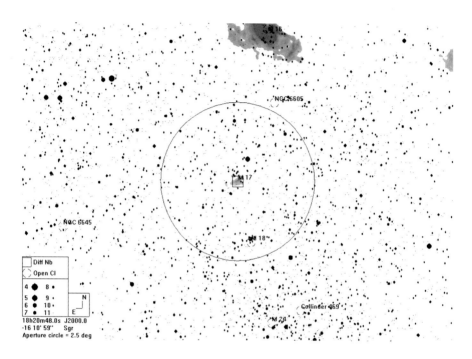

The Omega Nebula forms the southern apex of an equilateral triangle with γ Sct and **M16** (below) as its other apexes.

The initial impression is of an elongated glimmer of greyish light. Under examination with averted vision (or a UHC filter), an extension appears to the SW of the glimmer, giving the nebulosity the appearance of a tick (check mark) rather than an omega or a swan.

Sagittarius: Globular Cluster: M28 (NGC 6626) (70 mm)

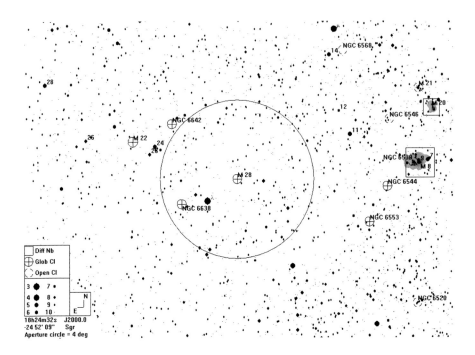

The magnitude 6.8 **M28** is in the same field of view as **λ Sgr**, one degree NW of the star.

M28 appears as a small, bright fuzzy disc, in which averted vision will bring out the core, which is almost stellar looking at this magnification.

Sagittarius: Open Cluster: M25 (IC 4725) (100 mm)

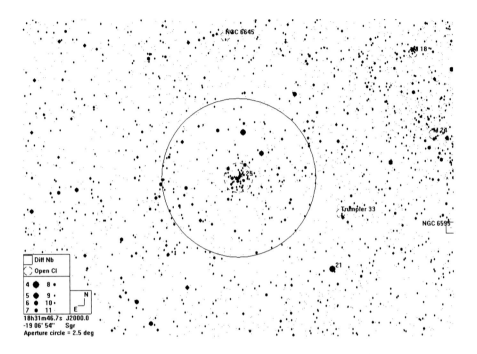

If you place γ **Sct** on the W edge of the field and scan S for 2 fields, you will find **M25** showing distinctly against the background Milky Way.

This bright rich cluster is unusual for open clusters in that it has few blue-white stars. Of the dozen or so stars that are resolved in big binoculars, note the triangle of deep yellow 7th magnitude stars to the N of the Cepheid variable, **U Sgr**, which has a mean magnitude range of 6.3–7.1 over a period of 6.7 days.

Sagittarius: Globular Cluster: M22 (NGC 6656) (70 mm)

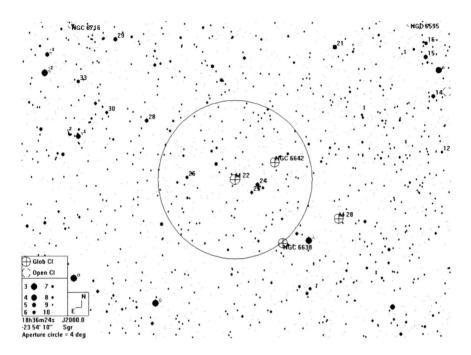

M22 is one field to the NE of **λ Sgr**, the "peak" of the "teapot lid" asterism.

This is a beautiful globular cluster at ×37, where it shows a very much brighter core, very much like the nucleus of a comet. This makes it very clear why Charles Messier compiled his catalogue of objects that were not to be confused with comets. It is the third largest of the southern hemisphere globular clusters and, despite this accolade usually being given to **M13** (which is easier to observe), is the largest globular that is visible from the UK.

Sagittarius: Globular Cluster: M54 (NGC 6715) (100 mm)

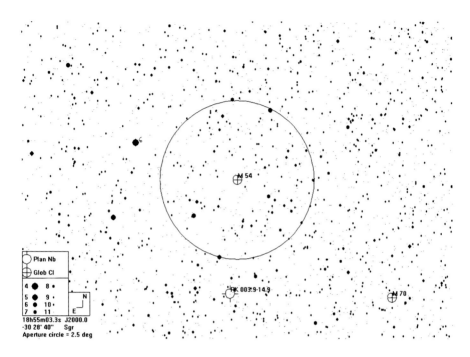

If you place *Ascella* (ζ **Sgr**) at the NE of the field of view, the magnitude 7.7 **M54** will be SW of center.

At 37×100, **M54** has the appearance of a small diffuse glowing ball. As with all globulars, it responds well to averted vision which, in this case, makes the almost stellar-looking core visible.

Sagittarius: Globular Cluster: NGC 6723 (100 mm)

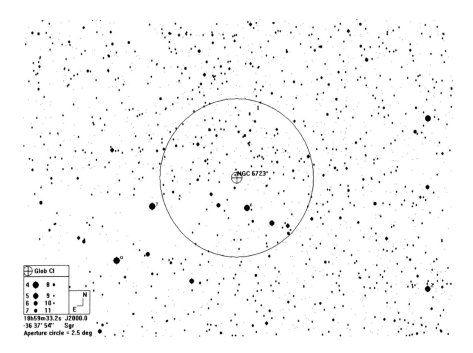

NGC 6723 is to the N of ε **CrA**, at the right-angled apex of a triangle that has its other apex at γ **CrA**.

With a diameter of 11 arcmin and a magnitude of 7, **NGC 6723** is another of those large bright southern globulars. It is fairly loose at its extremities and looks as though it is about to resolve at ×37.

Sagittarius: Globular Cluster: M55 (NGC 6809) (70 mm)

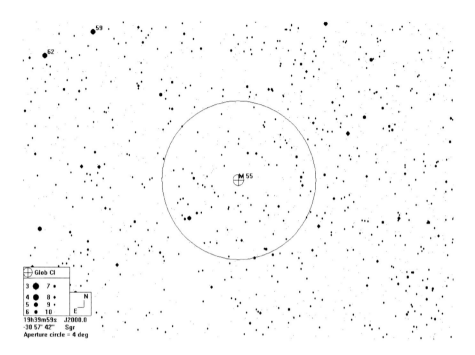

If you pan exactly 8° E of *Ascella* (ζ **Sgr**), the magnitude 6.3 **M55** should be near the center of the field of view.

M55 appears as a brightish glowing ball of nebulosity, which gets gradually but distinctly brighter towards the middle. With averted vision, it not only grows (as do almost all globulars), but it gains a slightly mottled appearance that makes it look as though it is almost ready to start resolving into stars.

Telescopium: Globular Cluster: NGC 6584 (100 mm)

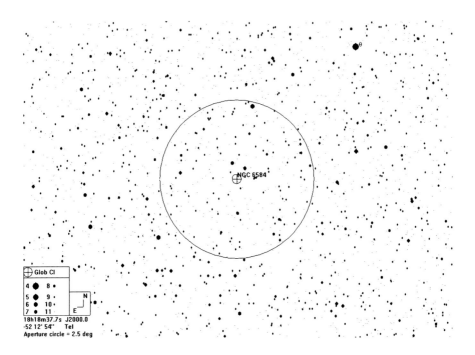

NGC 6584 is just over one field to the SE of the distinctively blue θ **Ara**.

NGC 6584 is a relatively small (8 arcmin), but distinctively bright globular cluster. It is entirely unresolved at ×37 but is one of those binocular objects that seems as though just a little more magnification will start to reveal its secrets.

Serpens: Emission Nebula and Cluster: M16 (NGC 6611, the *Eagle Nebula*) (100 mm)

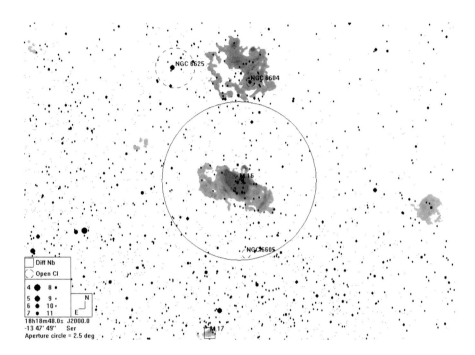

Identify γ **Sct**, place it at the S of the field of view, and pan one and a half fields to the W. The cluster associated with **M16** should be visible near the center of the field.

Unless skies are very good, you may only be able to see the cluster in an unfiltered view. A UHC filter will bring the nebulosity into prominence and you should be able to identify in its form the wings and tail from which it gets its common name.

Serpens: Open Cluster: IC 4756 (50 mm)

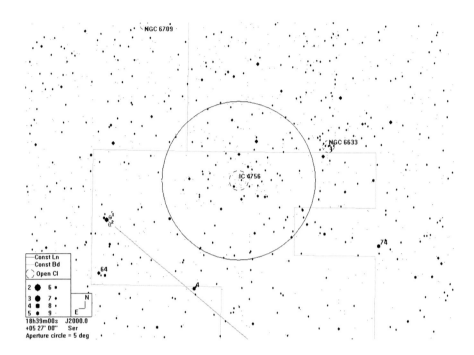

IC 4756 is 4½° ENE of θ **Serpentis**.

This delightful cluster, which is somewhat larger than the Moon in extent, has over a dozen members visible against a fuzzy backdrop in 10×50 binoculars. Also in the region is **NGC 6633**, which is smaller and less bright, but still has resolvable stars in medium-sized binoculars.

Serpens: Double Star: θ Serpentis (100 mm)

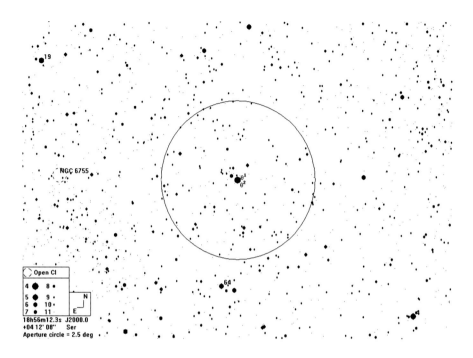

θ **Ser** is the tip of the tail of the snake. It is three fields to the W of δ **Aql**.

θ **Ser** is a pair of 5th magnitude stars of spectral type A5 separated by 22 arcsec. It is easily split at ×37 and, owing to the approximate equality of brightness of its components, is a good test of 10×50 binoculars.

Scutum: Open Cluster: M26 (NGC 6694) (70 mm)

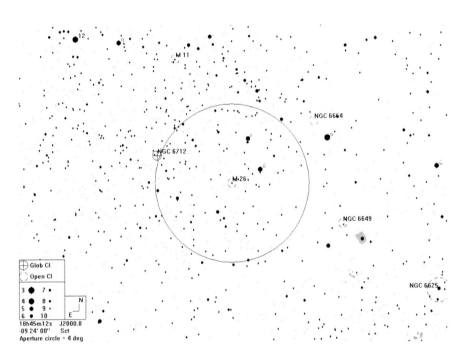

If you put α **Sct** at the NW of the field of view, **M26** will be SE of center.

Even at magnitude 8.0, **M26** is easy to find in a 42-mm binocular, being small (5 arcmin diameter) and condensed. In a 15 × 70, only one star is definitely resolved, with others forming a brighter glowing kite shape.

Scutum: Open Cluster: M11 (NGC 6705, *Wild Duck Cluster*) (50 mm)

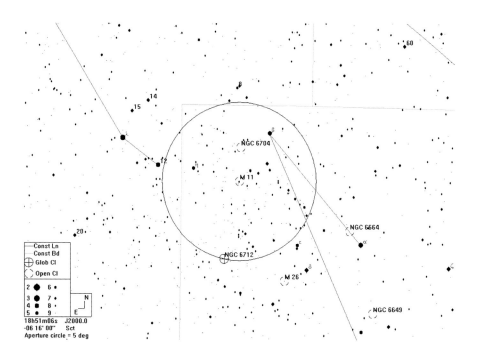

M11 is 4° to the ESE of the 3rd magnitude λ **Aquilae**.

In 10×50 binoculars, the cluster is seen as a bright wedge-shaped glow of light. Although they will not resolve the vee shape of brighter stars that gives this cluster of a thousand or so stars its common name, the cluster is still one of the better objects for this size of binoculars. Also worth enjoying in this region of sky is the denser part of the Milky Way that forms the *Scutum Star Cloud* as a backdrop to this cluster.

Scutum: Globular Cluster: NGC 6712 (100 mm)

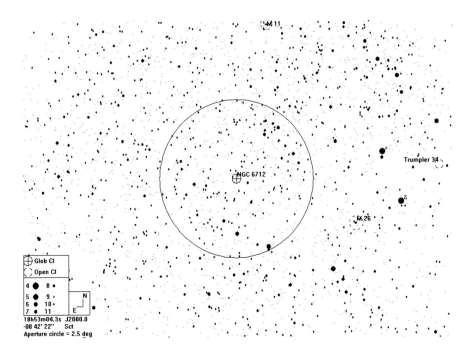

NGC 6712 is one 2.5° field E of **ε Sct**.

This 8th magnitude globular cluster is an easy object, about 5 arcmin in diameter. It is in a particularly beautiful star field. In particular note the various colors of the stars in the little equilateral triangle of 7th and 8th magnitude stars at the NW of the field.

Pavo: Globular Cluster: NGC 6752 (C 93) (100 mm)

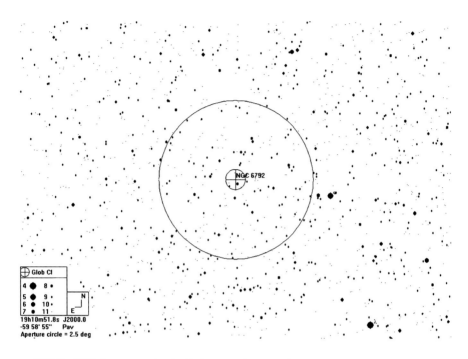

NGC 6752 is one and a half degrees from ω **Pav** in the direction of α **Pav**. For northern hemisphere observers, it may be easier to start at α **Pav**, from which you go four fields W and then one field S.

NGC 6752 is a large (19 arcmin) fairly loose globular that is one of the best of this class of object for large binoculars. It is bright (magnitude 5.4) and is worth looking for from any location south of the Tropic of Cancer.

Aquila: Open Cluster: NGC6709 (100 mm)

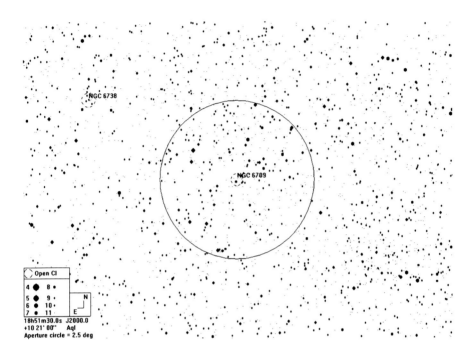

This magnitude 6.7 cluster lies 5° SW of ζ**Aql**.

It lies in a rich field of stars and is about half the diameter of the Moon. Although it is visible in smaller binoculars, it does not begin to resolve properly until the magnification increases.

Aquila: Open Cluster: NGC 6738 (100 mm)

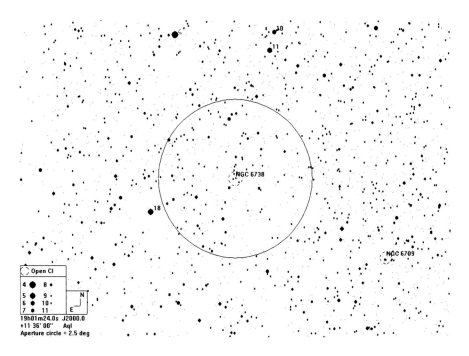

This magnitude 8.3 cluster lies 2.5° SW of ζ**Aql**.

This sparse cluster is about half the diameter of the Moon. A 100-mm binocular should resolve 20 stars of 9th and 10th magnitude. It is unusual in that it has no other stars brighter than 13th magnitude, and, for this reason, although it is classed as an open cluster, some suspect that it may be an asterism.

Aquila: Planetary Nebula: NGC 6781 (100 mm)

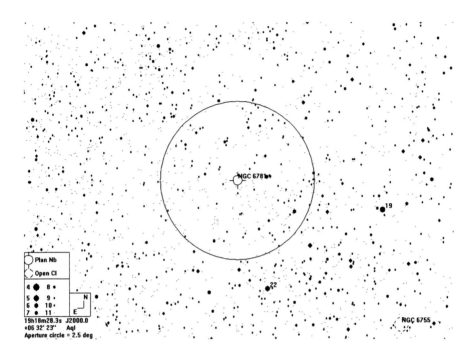

To locate **NGC 6781**, begin at **δ Aql** and hop 2 fields in the direction of **ζ Aql**. **NGC 6781** will be near the center of the second field.

This 12th magnitude planetary is possibly the most challenging object in this list and requires superb sky conditions and an experienced observer. Were its position not so easy to locate, it would have no place here at all. It is considerably easier to see and identify if you use a UHC or [O-III] filter. With such a filter and averted vision, it appears very slightly elongated at ×37.

Aquila: Dark Nebulae: Barnard 142, 143 (*Barnard's E*) (70 mm)

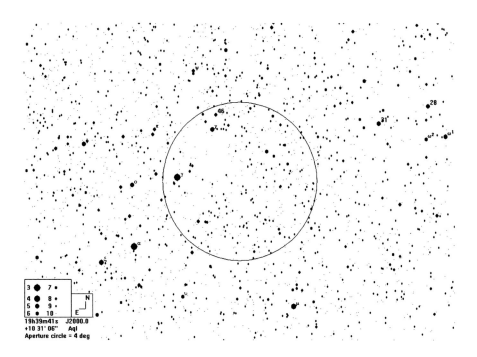

This distinctive pair of dark nebulae lies 3° NW of **Altair (αAql)**.

The pair lies in a rich star field that serves to make it more distinctive, where it appears to spell out either the letter "E" or an underlined "C".

Vulpecula: Asterism: (Cr 399, *Brocchi's Cluster*, the *Coathanger*) (50 mm)

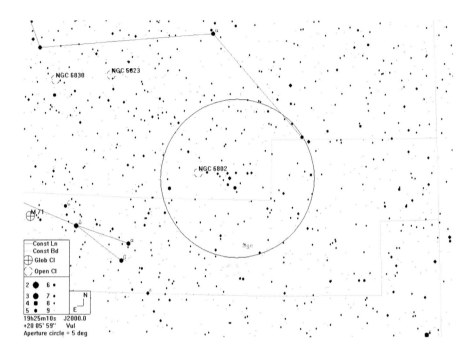

Cr 399 lies 5° S of α **Vulpeculae** and 4° NW of α **Sagittae**.

Even small binoculars will reveal the ten stars that give this asterism its common name. Because of this shape, the Coathanger makes a good star-party piece, with its 1½° span being neatly framed within a 5° field of view for sufficiently long periods to permit several consecutive observations in mounted binoculars.

Vulpecula: Planetary Nebula: M27 (NGC 6853, the *Dumbbell Nebula*) (50 mm)

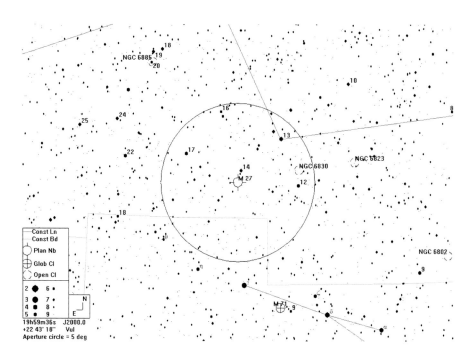

If you place γ **Sagittae** at the S of a 5° field of view, **M27** will be just N of center.

Although the *Dumbbell* is not the only planetary nebula that is visible in 10×50 binoculars, it is significantly easier to see than any other, but will need a larger instrument with more magnification to show some structure. The progenitor star is far too faint to be seen, even in large binoculars.

Sagitta: Double Star: ε Sagittae (100 mm)

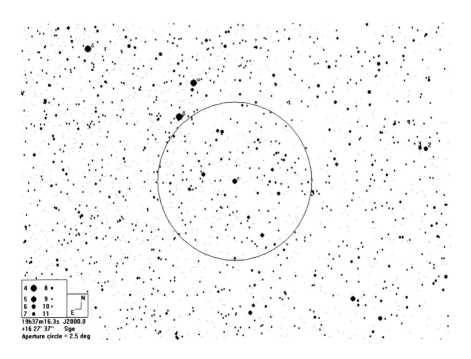

ε Sge is located about a degree and a half SW of **α** and **β Sge**, the fletch end of the arrow.

ε Sge is a beautiful colored pair with the 8th magnitude blue secondary 88 arcsec from the 6th magnitude yellow primary.

Sagitta: Cluster: M71 (NGC 6838) (100 mm)

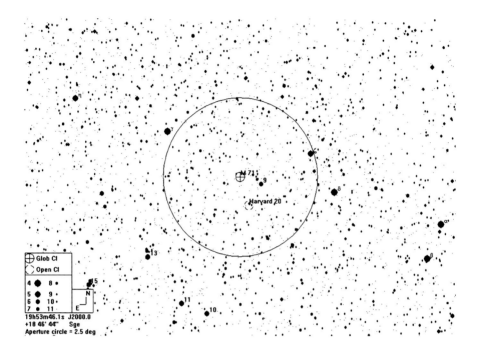

M71 is marginally south of the midpoint of a line joining γ and δ **Sge**.

It is easy to distinguish this small cluster from the background Milky Way in big binoculars at ×37. There is some dispute as to its nature, and it has been variously described as a compact open cluster and a loose globular cluster, with the latter having the most recent favorability. It is included because of this somewhat enigmatic nature, as it is otherwise a fairly banal object in binoculars.

Cygnus: Double Star: β Cyg (*Albireo*) (50 mm)

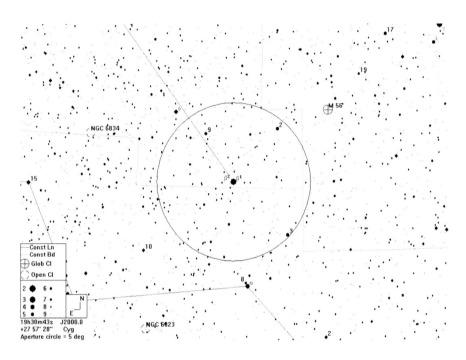

β Cygni is the star at the head of the Swan.

10×50 binoculars show both members of this superb double star, which are separated by 34 arcsec. The primary (3rd magnitude) is a deep orange (spectral type K) and the fainter (5th magnitude) comes is a bright sapphire-blue (spectral type B). The star is a true binary, with an orbital period of 7,270 years.

Cygnus: Open Cluster: M29 (NGC 6913) (70 mm)

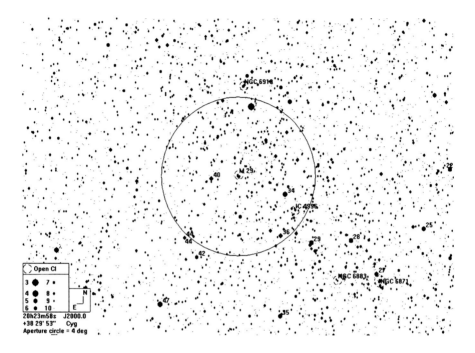

If you put **Sadr** (γ **Cyg**) at the N of the field, this magnitude 6.6 cluster is near the center, just under 2° S of the star.

This is a fairly sparse and unremarkable cluster in smaller binoculars, but, on a good night, a 15×70 will resolve around a dozen stars of the 50 or so in this small cluster. Some of the brighter stars appear to make the letter "H".

Cygnus: Dark Nebula: LDN 906 (B 348, the *Northern Coalsack*) (50 mm)

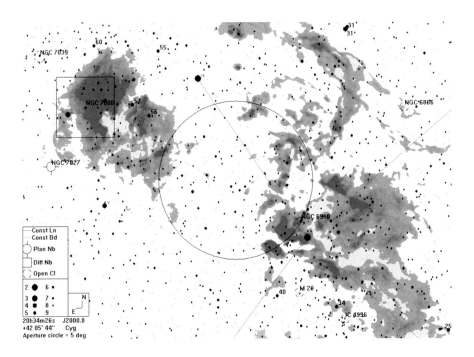

LDN 906 is the dark parch of sky which lies very slightly to the E of a line from *Deneb* (α **Cyg**) to *Sadr* (γ **Cyg**).

LDN 906 is one of several regions that are sometimes called "the *Northern Coalsack*". It is a region where interstellar dust obscures our view of the Milky Way.

Cygnus: Supernova Remnant: *Veil Nebula* NGC 6960 (C34), NGC 6992 (C33) and 6995 (100 mm)

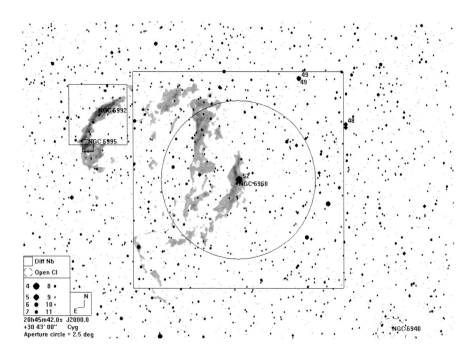

The western part of the *Veil* (**NGC 6960**) is a background to **52 Cyg** which, if the sky is dark enough to see the *Veil*, will be visible to the naked eye 3° S of **ε Cyg**. The *Eastern Veil* (**NGC 6992 and 6995**) can be found by scanning one field to the NE.

If you have a UHC or [O-III] filter, this object is a must. It is nothing less than superb in big binoculars in either of these filters. The first time I saw the eastern portion with an [O-III], my immediate reaction was that the binocular had *shrunk* the Milky Way! In a good sky, you will realize that even the field of the binoculars cannot contain the entire *Western Veil*, as small detached clumps of it come into view with averted vision.

Cygnus: Emission Nebula: NGC 7000 (C20, the *North American Nebula*) (50 mm)

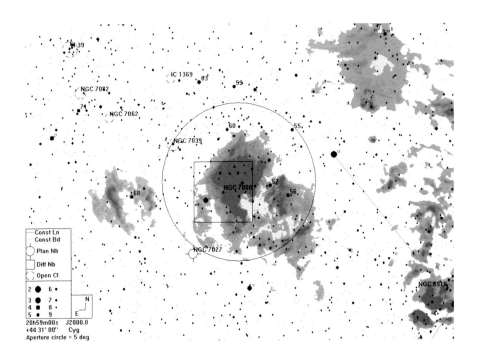

NGC 7000 is a bright patch of nebulosity whose center is about 4° ESE of α **Cygni** (*Deneb*).

In a transparent dark sky, it is visible to the naked eye with direct vision and easy with averted vision. It is extremely large, being about four times the diameter of the Moon, and is one of the few objects that is better in 7×50 binoculars than in 10×50. The characteristic shape is given to it by intervening clouds of dust.

Cygnus: Double Star: 61 Cygni (70 mm)

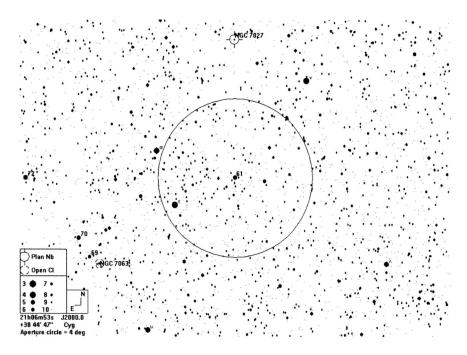

61 Cyg is situated to the WNW of τ **Cyg**, in the same field.

Every astronomer should observe **61 Cyg**! Not only was this star the first to have its distance measured (by F.W. Bessel in 1838), but it is also the naked eye star with the greatest proper motion (5.22 arcsec/year)—although some claim this for the star Groombridge 1830, which is at magnitude 6.4 and is visible from very dark sites (7.06 arcsec/year). It is at a distance of 11.4 light-years (Bessel measured it at 10.3).

61 Cyg has a combined magnitude of 4.8, but binoculars show it to be a pair of orange-red stars (both spectral type K) with magnitudes of 5.2 and 6.0. They have an orbital period of 659 years.

Cygnus: Open Cluster: M39 (NGC 7092) (70 mm)

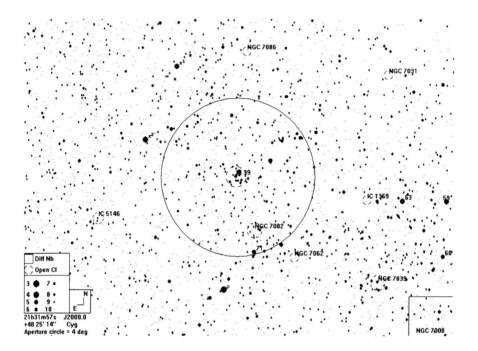

M39 lies in a rich part of the Milky Way, halfway between ***Deneb*** (**α Cyg**) and **α Lac**.

This is a relatively loose and bright (magnitude 4.6) cluster in which a 15×70 binocular will resolve around 20 of the 30 or so confirmed members of this cluster, which covers a bit more sky than the Moon. It is bounded by a triangle that has the cluster's brightest stars at its apexes.

Delphinus: Globular Cluster: NGC 6934 (C47) (100 mm)

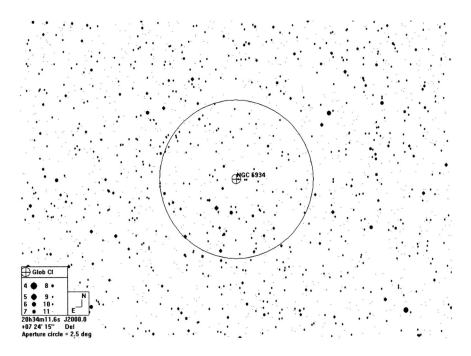

Scan two fields to the S of ε **Del**, and **NGC6934** will be obvious to the E of center of the second field.

NGC 6934 is a very easy object, even in 8×30 binoculars. There is a phenomenon that is common to almost all largeish globular clusters in 100-mm glasses at ×37: they appear to get larger as transparency improves. **NGC 6943** displays this phenomenon in a more pronounced degree than any other globular which I have tested for it.

Pegasus: Globular Cluster: M15 (NGC 7078) (50 mm)

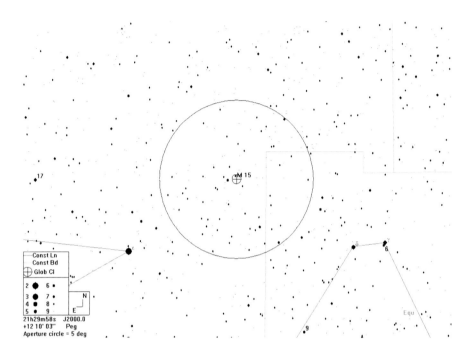

M15 is a little over 4° to the NW of the 2nd magnitude ε **Pegasi** (*Enif*)

Although M15 is small, with about half the apparent size of the better-known M13, it is very bright and for this reason is one of the better globular clusters for small binoculars. While you are observing in that region, go back to ε **Peg**, which is a binocular double; the 8.6th magnitude secondary is 2.4 arcmin from the yellowish primary, in the direction of M15.

Aquarius: Globular Cluster: M2 (NGC 7089) (50 mm)

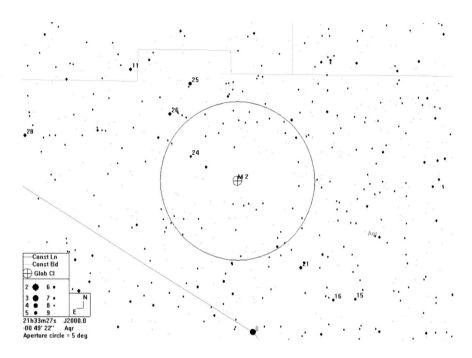

This magnitude 6.5 globular cluster is about 5° N of the 3rd magnitude β **Aquarii** (*Sadalsud*).

M2 is small and bright, having the appearance of a fuzzy star in 10×50 binoculars, but is visible in less than ideal sky conditions. At 50,000 light-years, it is at a greater distance from us than either **M13** or **M5** and has a diameter of about 150 ly.[1]

---

[1]Burnham 1978, p. 188

Aquarius: Double Star: Struve 2809 (100 mm)

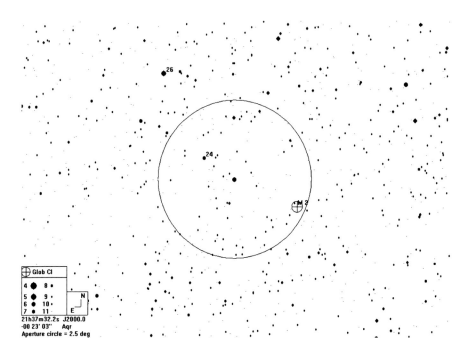

First locate the globular cluster **M2**, which is about 5° N of the 3rd magnitude β **Aqr** (*Sadalsuud*). **Struve 2809** is the nearer of the two 6th magnitude stars that form a line to the NE.

This is a close pairing (31 arcsec) which can be a challenge in smaller binoculars. The stars are 6th and 9th magnitude, respectively.

Cepheus: Open Cluster: IC1396 (50 mm)

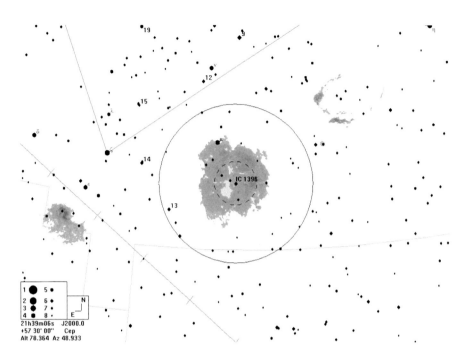

IC 1396 is a degree and a half S of the *Garnet Star* (μCep)

The magnitude 3.5 **IC 1396** is an extremely large cluster, over a degree and a half in diameter. It is elongated along a NE-SW axis. It is just possible to make out the surrounding nebulosity on a dark, transparent night, if you hold a UHC filter in front of one of the eyepieces.

Cepheus: Red Giant: μ Cep (the *Garnet Star*) (50 mm)

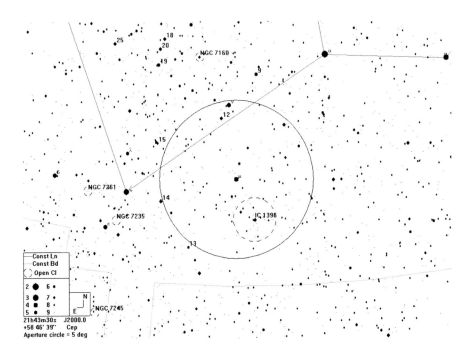

To find the *Garnet Star*, place α **Cephei** at the NW edge of a 5° field and μ will be diametrically opposite.

μ Cep is one of the reddest stars in the sky; it was named "the Garnet Star" by William Herschel. The deep orange color of this red giant is nicely brought out in 10×50 binoculars. It has a variability of a bit less than a magnitude but is usually around 4th magnitude. It is one of the largest known stars; if it replaced the Sun, it would extend well beyond the orbit of Jupiter. It is destined to become a supernova.

Chapter 14

September Equinox to December Solstice (RA 22:00 h to 04:00 h)

Lacerta: Open Cluster: NGC 7209 (70 mm)

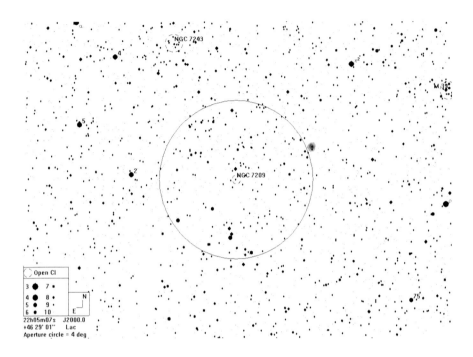

NGC 7209 is just under 3° W of the 4.5th magnitude **2 Lac**.
In 15×70 binoculars, this tight (15 arcmin diameter) magnitude 7.7 cluster shows six stars in a sort of "omega" configuration, resolved against the background fuzziness of unresolved stars. There are representatives of each magnitude cohort of stars, so increasing aperture and magnification will show progressively more stars: I can only resolve two stars with a 10×42, but 31 are resolved with a 37×100.

Lacerta: Open Cluster: NGC 7243 (70 mm)

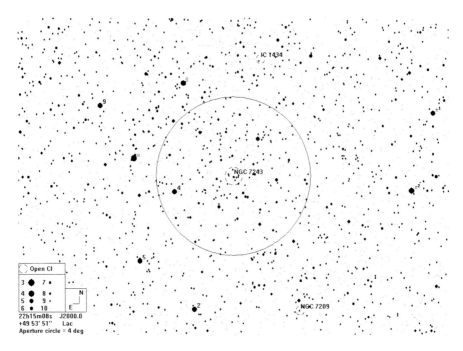

NGC 7243 is just over a degree and a half west of the magnitude 4.5 star, **4 Lac**.

NGC 7243 is much bigger (25 arcmin), brighter (magnitude 6.4), and looser than the other Lacerta cluster, **NGC 7209**, and is a fine object in binoculars of any size. My 10×50 resolves six stars and a hint of structure, but the 15×70 sees 20 or so and distinct dark lanes that give a sense of a triangular shape to the cluster. Over 50 stars are resolved in the 37×100.

Cepheus: Open Cluster: NGC 7235 (70 mm)

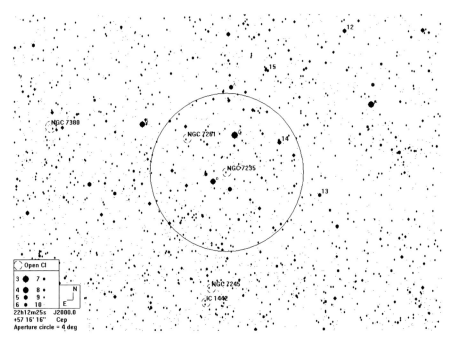

NGC 7235 is a degree south of ζ **Cep**.

This tight (6 arcmin diameter) triangular magnitude 7.7 cluster is a nice object in anything from 70-mm binoculars upward, in which the color difference between the two brightest stars, which are just on the east side of the cluster, just begins to become apparent in a sufficiently dark sky; I find that the difference becomes slightly more apparent if I gaze at it for a minute or so. The background fuzziness of fainter stars becomes more apparent with averted vision.

Cepheus: Open Cluster: NGC 7510 (70 mm)

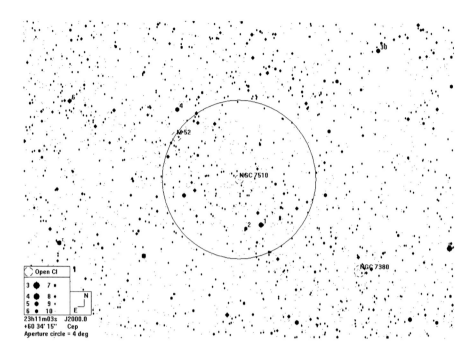

NGC 7510 is 2° southwest of **M52**

NGC 7510 is a tiny (4 arcmin) but very obvious cluster. No stars are resolved, but it appears distinctly elongated (about 2:1) in a 15×70 binocular. It is thought to be about 10 million years old.

Aquarius: Planetary Nebula: NGC 7293 (C63, the *Helix Nebula*) (100 mm)

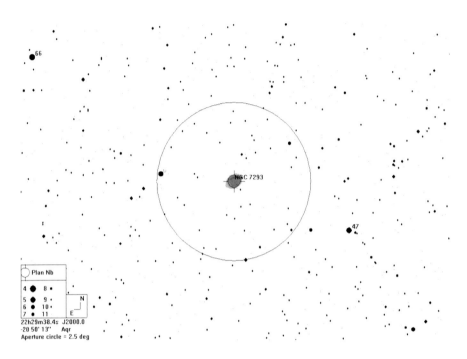

Identify the 3.6th magnitude **88 Aqr** and place it on the S edge of the field or the finder ring. Sweep four fields (9.25°) to the W and **NGC 7293** should appear in the eyepiece.

With an integrated magnitude of 6.5, the *Helix* seems as if it should be an easy object. However, it is a large (about the size of the Moon) object with low surface brightness and it requires a dark sky or a UHC or [O-III] filter. In 100-mm binoculars I find it slightly easier at ×20 than at ×37. The middle of this nebula is noticeably darker than the periphery at both magnifications, particularly with a filter.

Sculptor: Galaxy: NGC 55 (C72) (100 mm)

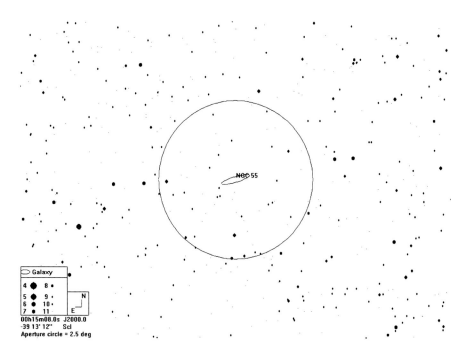

NGC 55 is located one and a half fields to the NW of α **Phe**.

This Sculptor Galaxy is not visible from the latitude of Britain. It is slightly dimmer than **NGC 253**, but is noticeably longer and thinner. It is neatly framed in a 2.5° field, making it a very nice binocular object.

Sculptor: Galaxy and Globular Cluster : NGC 253 (C65) and NGC 288 (70 mm)

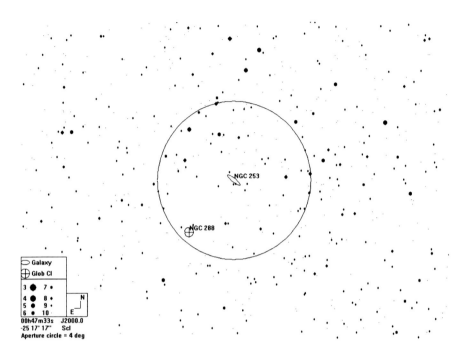

5° S of *Diphda* (β **Cet**) find a triangle of 5th magnitude stars. Place the most southerly star at the N of your field of view, and **NGC 253** is near the S edge of the field.

This bright galaxy shows as an elongated glow with a brighter middle. It is a relatively easy object, even from the latitude of Britain, despite to its low transit altitude. It is so bright and large (over 20 arcmin long) that it is possible to find and identify, even in smaller binoculars. A good southern horizon is, of course, essential from this latitude.

The globular cluster **NGC 288,** which lies in the same field, to the SE, is another easy object, showing as a dim circular glow with about half the diameter of the galaxy's length.

Sculptor: Galaxy: NGC 300 (C70) (100 mm)

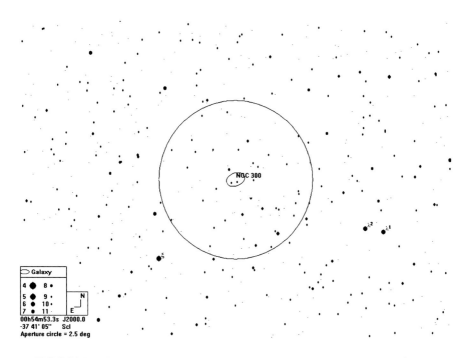

NGC 300 makes the slightly obtuse apex of an isosceles triangle with ξ **Scl** and λ^2 **Scl** as the base angles.

Imagine a mini-version of **M33** and you have **NGC 300**. It is to the 100-mm instrument what the Messier galaxy is to a 50-mm binocular. It is apparently 9th magnitude, but is of extremely low surface brightness. It is difficult from northern temperate latitudes and is far better seen from the tropics or southern latitudes.

Vela: Open Cluster: NGC 3228 (100 mm)

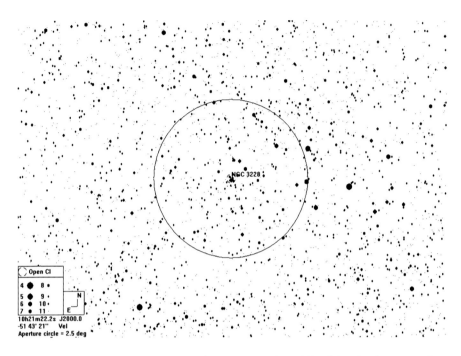

NGC 3228 is situated about half a degree to the NW of the dead center of an imaginary line joining ϕ **Vel** and μ **Vel**.

NGC 3228 is a small (5 arcmin) cluster of mainly white stars, of which nine are easily visible. It is a southern hemisphere version of the sort of binocular cluster that is in great abundance in the Perseus-Cassiopeia region of the northern hemisphere.

Tucana: Globular Cluster: NGC 104 (C106, 47 Tucanae) (100 mm)

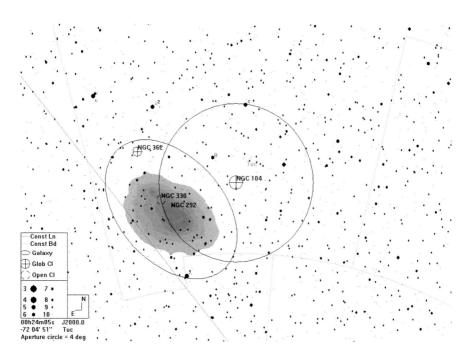

47 Tucanae is an easy naked eye object about a degree to the NW of the ***Small Magellanic Cloud (SMC)***.

This is an absolutely superb object in binoculars of any size. Big binoculars begin to resolve its outer regions. Although it is not quite as large or bright as its rival, ω ***Cen***, I find that it seems to resolve a bit better.

Also visible in binoculars is the otherwise fine, somewhat less impressive, globular, **NGC 362 (C104)**, which lies on the N edge of the *SMC*.

Tucana: Galaxy: NGC 292 (*Small Magellanic Cloud*) (50 mm)

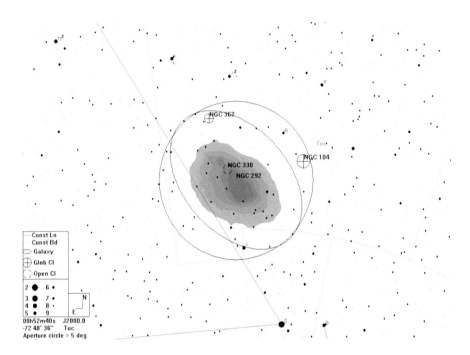

The *Small Magellanic Cloud* is easily visible to the naked eye just over 4½° NNE of β **Hydri**.

With a diameter of about 3°, the *Small Magellanic Cloud* is neatly framed in a 5° field of view. It is one of the satellites to our own galaxy, at a distance of 190,000 light-years. It was through studies of *Cepheid Variables* in the *Small Magellanic Cloud* that Henrietta Leavitt established their period-luminosity relationship, thus enabling their use as standard candles for measuring distances. Also visible to the naked eye is the bright globular cluster *47 Tucanae*, which is described among the 100-mm objects.

Andromeda: Galaxy: M31 (NGC 224, the *Great Andromeda Galaxy*) (50 mm)

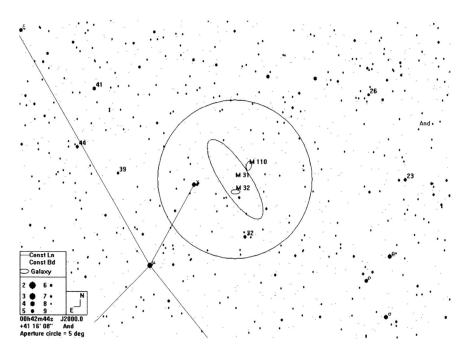

This magnitude 4.3 galaxy, which can be visible to the naked eye, is an easy star-hop from the yellowish β **And** *(Mirach)*. Place β near the SE edge of the field and find μ to the NW. Place μ where β was, and M31 will lie where μ was.

You should be able to see the elongated shape of M31 which, with patience and dark skies, extend almost across the field of view. Notice the significantly brighter glow of the nucleus and how the light of the galaxy drops off more abruptly at the NW edge as a consequence of a dust lane.

If you have good skies (or larger binoculars), you may be able to find the two companion galaxies. To the S of the nucleus lies M32 (NGC 221), making a right-angled triangle with two 7th magnitude stars and appearing like a large, slightly fuzzy, star in 10×50 binoculars. You may need to use averted vision to see this. Slightly more obvious in 10×50s, and to the NW, is M110 (NGC 205).

Andromeda: Open Cluster and Double Star: NGC 752 (C28) and 56 And (70 mm)

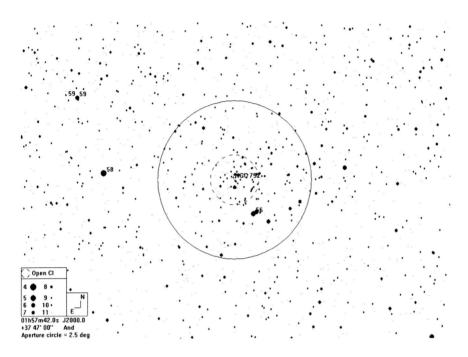

From **β Tri** hop 3° N to **58 And**, and then 2° W to **NGC 752**, which is just NW of the double star, **56 And**.

Although this cluster is visible in smaller binoculars and is often included in lists for them, it is significantly better in larger instruments. Several tens of stars (depending on sky conditions) become visible with 37×100 binoculars.

56 And is a beautiful air of 6th magnitude deep yellow stars separated by about 3 arcmin.

Cetus: Galaxy: NGC 247 (C62) (100 mm)

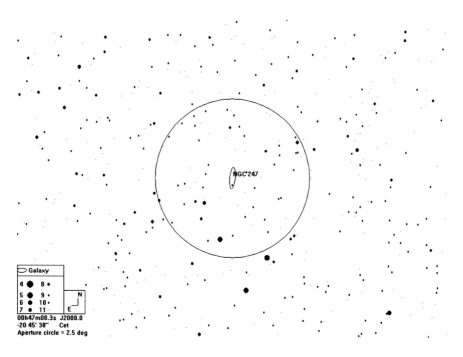

NGC 247 is nearly 3° SSE of **β Cet** (*Diphda*) and 4.5° N of the considerably easier **NGC 253**.

I find **NGC 247** to be extremely challenging in binoculars and will not even attempt it unless sky conditions are extremely good and I am relatively fresh. It is said to have a magnitude of 9.6 but, owing to its low surface brightness, it appears to be considerably fainter. The triangle of 5.5th magnitude stars just to the south of it makes it easy to be certain of its location. Once there, relax, use averted vision, and, if necessary, tap the binoculars to jiggle them slightly. This is usually sufficient to tease it from the background, but I am sometimes not sure if I have *really* seen it.

Pisces: Double Star: ψ^1 Piscium (100 mm)

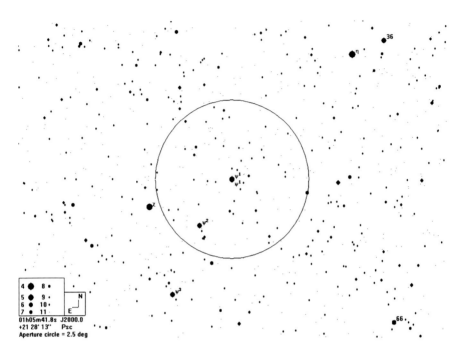

ψ^1 **Psc** lies in the northern branch of the Pisces asterism, a degree and a half from χ **Psc** in the direction of η **Psc**.

This is a delightful double of two brilliant white 5th magnitude stars separated by 30 arcsec.

Pisces: Double Star: ζ Piscium (100 mm)

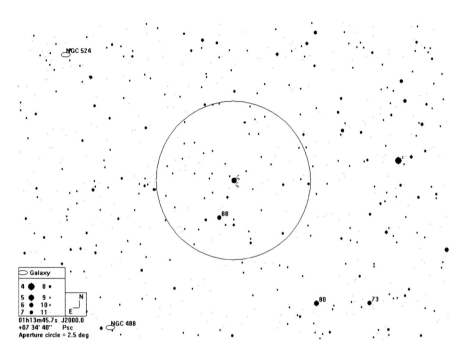

ζ **Psc** lies on the southern "branch" of the constellation, just over one field E of the brighter ε **Psc**.

This is a very pretty pair of 5th and 6th magnitude separated by 23 arcsec.

Andromeda: Open Cluster: NGC 7686 (70 mm)

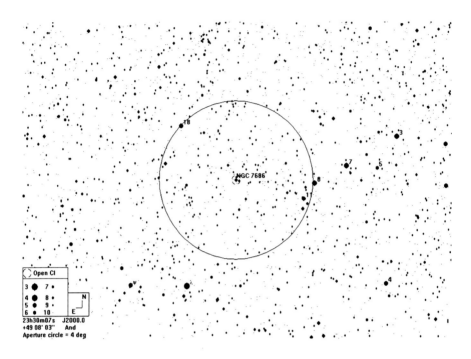

NGC 7686 is located 3° northwest of the 4th magnitude λ **And**.

This magnitude 5.6 cluster, which has about half the diameter of the Moon, can be visible to the unaided eye on a dark, transparent night. It has one very bright (magnitude 6.3) star in its center and then one or more stars in each magnitude band down to 11th magnitude, eight of which are visible in 15×70 binoculars.

Cassiopeia: Open Cluster: Stock 12 (70 mm)

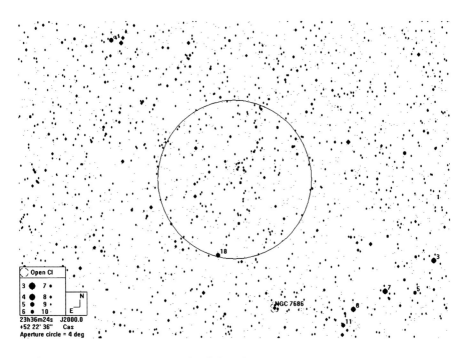

Stock 12 is nearly 4° north of **NGC 7686**

Although it is large (25 arcmin) and bright (a dozen stars of 8th–10th magnitude), **Stock 12** is not an obvious cluster and can be difficult to distinguish from field stars in small binoculars. As such, it can offer a challenge of identification. It does not help that a number of sources give the coordinates of a location about a third of a degree to the north, leading a number of observers to comment that they couldn't find **Stock 12**, but that there is a beautiful cluster just to the south!

Cassiopeia: Open Cluster: M52 (NGC 7654) (100 mm)

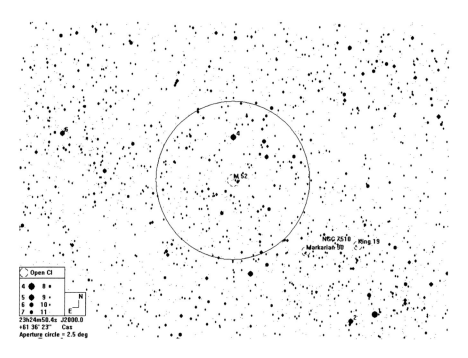

M52 is most easily found by continuing a line from **Schedar** (αCas) to **Caph** (βCas) for slightly more than the same distance again.

M52 is situated in a less dense region of the Milky Way, making it easy to spot. It has a triangular shape, and a 100 mm will show the distinctively yellower color of the brightest (8th magnitude) star near the western side of the cluster, contrasting with the blue-white of the other cluster stars. This distinction is because this brighter star is a foreground star, not a member of the 5,000 light-year distant magnitude 6.9 cluster, whose brightest star is magnitude 11.

Cassiopeia: Open Cluster: NGC 7789 (70 mm)

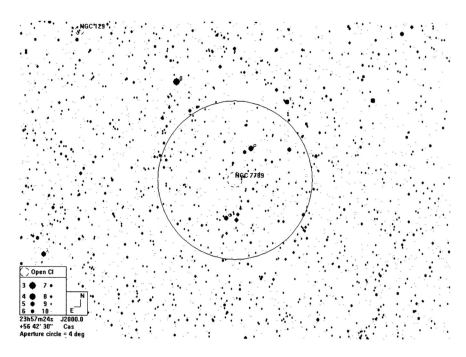

NGC 7789 is midway between ρ and σ Cas. If the sky is too bright for these stars to be confidently identified with the unaided eye, start at *Caph* (β Cas) and find ρ Cas 2.5° to the southwest. Just be careful that you do not misidentify τ Cas as ρ Cas and end up being unable to find the cluster between τ and ρ—it is easily done! ρ is a wide double, with distinctly yellow and blue components.

NGC 7789 is a large (16 arcmin), very rich cluster. At magnitude 6.7, it is obvious even in 10×50 binoculars, but a 15×70 starts to show a hint of some structure as the richer parts of it shine a little brighter than the less dense parts, giving it an almost cometary appearance. Although I am unable to confidently resolve any stars with the 70-mm binocular, with averted vision bits of it seem to twinkle and I suspect six or seven individual stars against the fuzzy glow. NGC7789 is 6,000 light-years away and thought to be 1.6 billion years old.

ρ Cas is an interesting object in its own right. It is a yellow hypergiant and is one of the most luminous stars known

Cassiopeia: Open Cluster: NGC 225 (70 mm)

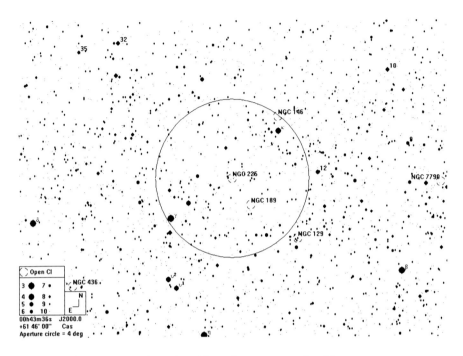

NGC 225 lies approximately midway between γ and κ **Cas**.

NGC 225 is one of those Milky Way clusters that can initially be mistaken for just a denser part of the Milky Way itself. However, it is a distinct cluster with 70-mm binoculars in which eight or nine stars will be resolved from the general glow. The tiny cluster half a degree to the west is **Stock 24**.

Cassiopeia: Open Cluster: NGC 436 (100 mm)

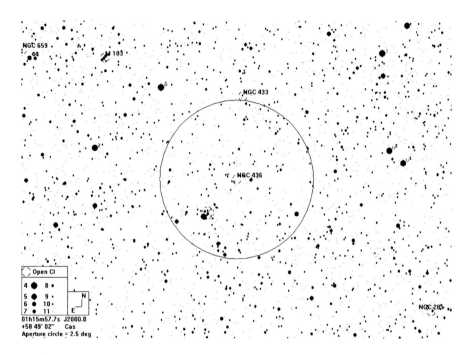

To find **NGC 436**, extend a line from **Segin (εCas)** to **Ruchbah (δCas)** a further 2°.

This 9th magnitude cluster is small (only about 6 arcmin across) but bright for its size. It is of interest mainly because it is a convenient stepping-stone to the far more interesting *Owl Cluster* (**NGC 457**), which lies just under a degree to the SE, so is worth recognizing.

Cassiopeia: Open Cluster: NGC 457 (C13) (the *ET Cluster*, the *Owl Cluster*) (100 mm)

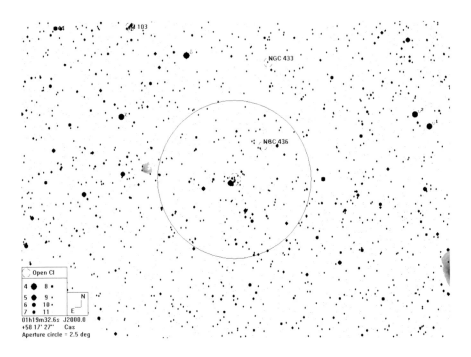

Place δ **Cas** (*Ruchbah*) at the NE of the field and the 5th magnitude φ **Cas** will lie diametrically opposite. Center φ **Cas**, which is the brightest star of this cluster, although it is not actually a member of it.

NGC 457 is near the top of my list of star-party objects. The two brightest stars, φ **Cas** and its nearby 7th magnitude companion, appear as the glowing eyes at the SE of a stick-man asterism with outstretched arms, which, just detectable at ×37, gives this cluster one of its common names.

Cassiopeia: Open Cluster: NGC 663 (C10) (50 mm)

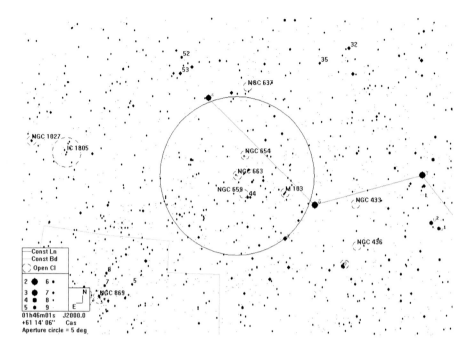

NGC 663 is the largest of several open clusters that are visible in the same 5° field as **δ** and **ε Cassiopeiae**.

It is superior in every respect to the nearby M103 and, unlike M103 which appears as a small fan-shaped patch of nebulosity, some of the stars are resolvable in 10×50 binoculars.

Cassiopeia: Open Cluster: NGC 654 (70 mm)

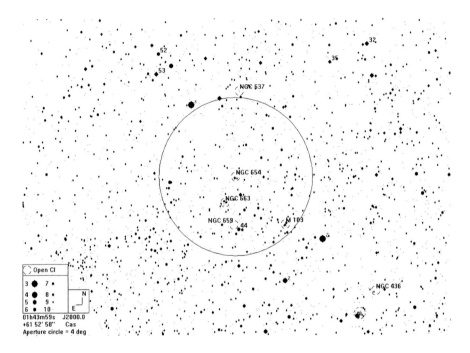

NGC 654 is just over halfway from δ to ε **Cas**; it is in the same field of view as **NGC 663**.

NGC 654 appears as the more northerly apex of a tight equilateral triangle that has a bright (magnitude 7) star and faint (magnitude 9) star as the other apex; the brighter of these is over 200 times more luminous than the Sun. **NGC 654** appears as a small, but very distinct, misty glow. It is about 20 million years old and 7,000 light-years distant.

Cassiopeia: Open Cluster: Cr 463 (70 mm)

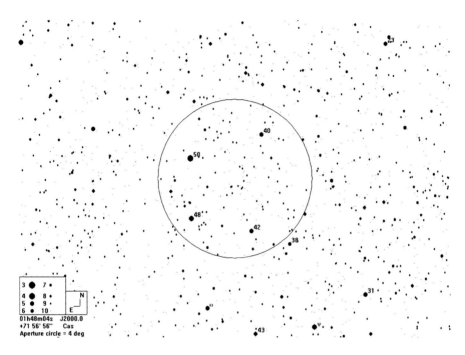

Cr 463 is located two thirds of the way from **Polaris (*a* UMi)** to *e* **Cas.**

At magnitude 5.7, **Cr 463** is a distinct object even in 10×50 binoculars, in which 6–8 stars are resolved, depending on your local conditions. In 70-mm binoculars, the number of resolved stars doubles. It is about 2,000 light-years away and 150 million years old.

Cassiopeia: Open Clusters: Mel 15 and NGC 1027 (70 mm)

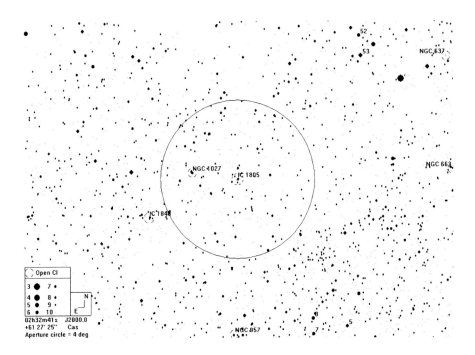

Melotte 15 is slightly over halfway from δ **Cas** to the 4th magnitude **CS Cam**. It will just fit in the same 15×70 field as **Stock 2**, if you hold the latter at the SW edge of the field of view.

Mel 15 is the cluster that is associated with the nebulosity **IC 1805** (the **Heart Nebula**). The nebulosity is very faint and, although it is more easily visible in 100-mm binoculars with a UHC filter, the more contrasty streak of it that extends to its south can be visible in 70-mm binoculars with a UHC filter on a very transparent night. The cluster itself, which is about double the apparent diameter of the Moon, can be identified by a cruciform asterism of slightly brighter stars.

The smaller (20 arcmin diameter) **NGC 1027** is a degree to the east. You should be able to resolve six stars with a 70-mm binocular.

Camelopardalis: Open Cluster: Stock 23 (70 mm)

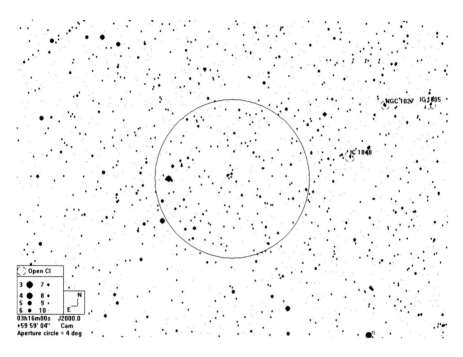

Stock 23 is a degree and a half due west of **CS Cam**.

Stock 23 is an obvious object in 70-mm binoculars, where it appears to have the form of a "mini-Pleiades" about 15 arcmin in extent, with a dozen or so stars resolved in a 15×70 binocular.

Andromeda: Open Cluster: NGC 956 (100 mm)

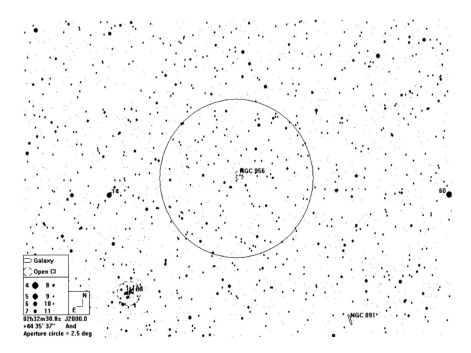

NGC 956 is 2.5° NW of **M34**.

NGC 956 is a small, faint (mag 8.9) cluster which is characterized by three obviously brighter stars that stand out against the misty background glow of the rest of the cluster. The brightest of these appears distinctly yellower than the other two, a distinction that increases the longer you look at it.

Triangulum: Galaxy: M33 (NGC 598, the *Pinwheel Galaxy*) (50 mm)

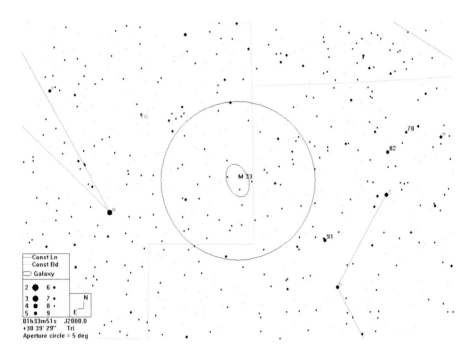

M33 is located a little over 4° from α **Trianguli** in the direction of β **Andromedae**.

M33, which has a high integrated magnitude (and can be visible to the naked eye under ideal conditions), is a large object with a low surface brightness. It therefore requires a dark sky and low magnification, making it easier to find and see in 10×50 binoculars than in small telescopes.

Aries: Triple Star: 14 Arietis (50 mm)

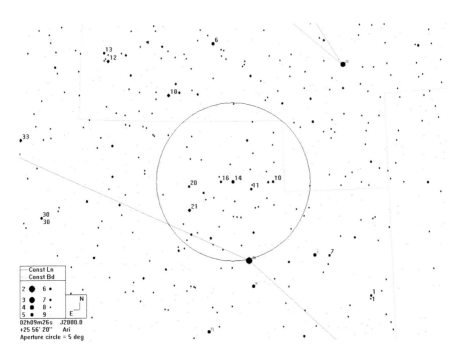

14 Arietis is 2.5° N of **Hamal** (α **Ari**)

The brighter two members (5th and 8th magnitude) are easy and separated by 103 arcsec. The third member is considerably fainter (11th magnitude) and is 10 arcsec closer to the primary. It is thus a challenge in 50-mm binoculars.

Eridanus: Galaxy: NGC 1232 (100 mm)

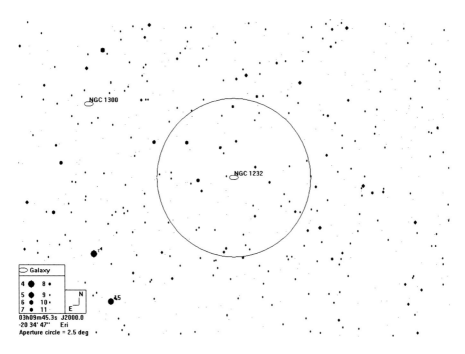

NGC 1232 is 2½° NW of τ^4 **Eri**. It is immediately adjacent to a 9th magnitude star (**HD 19764**) that lies 7 arcmin to the east of it.

The magnitude 9.8 **NGC 1232** is a low surface-brightness object which benefits from the lower magnification of binoculars. If it is not immediately obvious (it can look stellar with direct vision), make sure that you have **HD 19764** centered in the field of view and then use averted vision, which should make the galaxy appear. It is one of those objects that is very amenable to "playing" with averted vision, brightening and growing, and then shrinking again, depending on how you look at it.

Cetus: Variable Star: o Ceti (*Mira*) (50 mm)

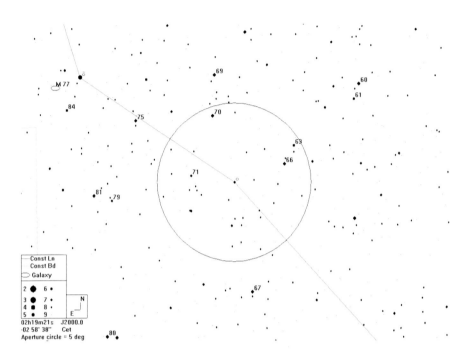

Mean magnitude range: 3.6–9.3
Mean period: 334 days
Type: Mira

Mira can be traced throughout most of its period with medium-sized binoculars. It is usually invisible to the naked eye, but its peak magnitude ranges from about 4th magnitude to brighter than 2nd magnitude (William Herschel noted that the 1799 maximum rivaled *Aldebaran* in brightness[1]). *Mira* has been known to be variable since 1596, prior to the first astronomical use of the telescope. It gives its name to a class of variable red giant stars that share the same characteristic variability.

---

[1] Levy, p. 48

Cetus: Galaxy: M77 (NGC 1068) (100 mm)

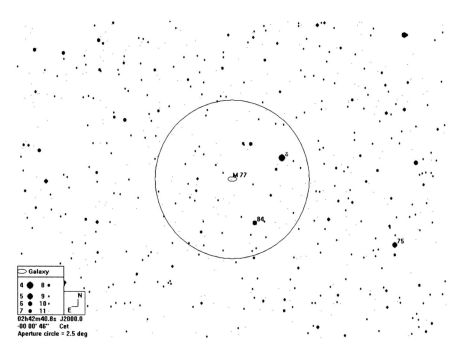

The magnitude 8.9 **M77** is less than a degree to the E of **δ Cet**.

This is a compact, nearly round, galaxy that could easily be confused for a globular cluster. It's worth learning to find this galaxy quickly if you ever intend to do a Messier Marathon, as it is one of the first objects you need to observe as the skies darken.

Cassiopeia: Open Cluster: Stock 2 (the *Muscleman Cluster*) (70 mm)

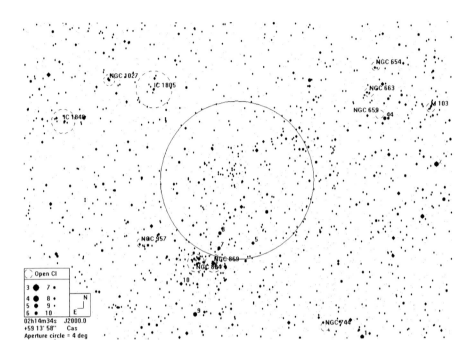

You can find **Stock 2** by going 2° N of the **Perseus Double Cluster**.

I discovered this magnitude 4.4 cluster by accident, during a less than competent attempt to locate the **Perseus Double Cluster** from light-polluted school playground. I was immediately struck by its delicate beauty: tens of 9th and 10th magnitude stars spread reasonably evenly over an area about a degree in diameter. It was only after I identified it and subsequently read other people's accounts that I became aware of the decapitated stickman, flexing his muscles and hauling a string of stars away from the **Double Cluster**, that give it its common name.

Although it looks larger than the Double Cluster, with a true span of 18 light-years, it is actually less than a third of the diameter of either component of the latter. It is much closer, only 1,050 light-years away, as compared to the 7,200 and 7,500 light-year distances of the two components of the better-known neighbor.

It is easily visible in 10×50 glasses, but I far prefer the view in 15×70s, where it shows more stars and is nicely framed by the smaller field of view.

Perseus: Open Clusters: NGC 884 and NGC 869 (C14, the *Double Cluster*) (50 mm)

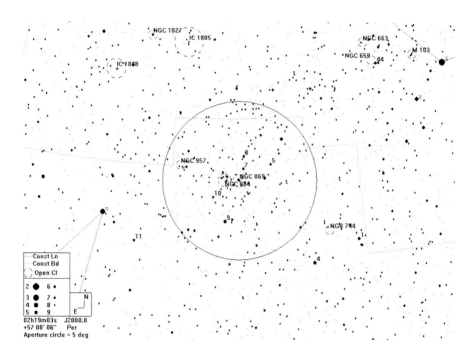

The clusters (magnitudes 5.1 and 5.3) are often visible to the naked eye, but, if not, follow a line from γ to δ **Cassiopeiae** for a distance beyond γ **Cas** to 1½ times the distance between the stars.

This pair of clusters in the sword handle of Perseus is a superb target, even in relatively small binoculars. There are several orange-red stars and these, combined with the varying brightness of the other stars, contrive to give it a three-dimensional appearance. As an aside, if our Sun was at the same distance as the *Double Cluster* (7,200 and 7,500 light-years, respectively), it would be too faint to be seen in 10×50 binoculars. It is worth scanning the region around the *Double Cluster*, as it is very rich in open clusters. The chain of stars to the NNW, for example, leads to **Stock 2**, otherwise known as the *Muscleman Cluster*, on the border of Perseus and Cassiopeia.

Perseus: Open Cluster: M34 (NGC 1039) (50 mm)

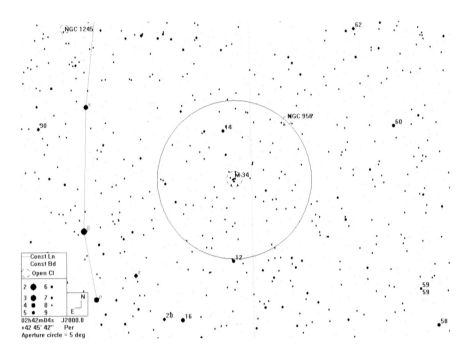

M34 is a degree NE of the midpoint of a line joining β **Persei** (*Algol*) to γ **Andromedae** (*Almaak*). It is visible to the naked eye under ideal sky conditions.

This is a superb cluster for binoculars of any size, showing a dozen or more stars in a 10×50. When you observe in this region, you could also ascertain the magnitude of the β **Per**, one of the more famous variable stars; it is an eclipsing binary star. The name *Algol* comes from the Arabic *Ras al Ghul*, the *Demon's Head*. The ghoul or demon in question is Medusa.

Perseus: Open Cluster: Melotte 20 (Cr 39, the *Alpha Persei Moving Cluster*) (50 mm)

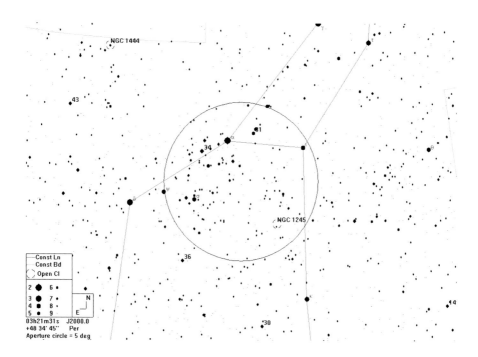

Melotte 20 is the cluster of stars, many of which are visible to the naked eye, around α **Persei** (*Mirfak*). These stars are mostly very bluish (spectral types O and B), indicating their relative youth (a few tens of millions of years). This is an ideal object for small and medium binoculars, which are able to encompass most or all of the cluster in their field of view.

Perseus: Open Cluster: NGC 1342 (70 mm)

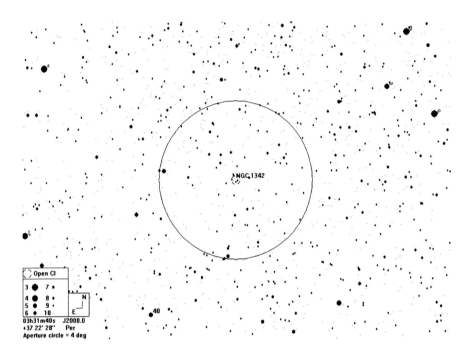

This cluster is best located with the aid either of a reflex finder or with wider-field binoculars. If you aim at a spot halfway between *Algol* (β **Per**) and ζ **Per**, **NGC1342** will be in the field, slightly towards *Algol*.

NGC 1342 is a sparse cluster of some 40 stars, most of which form a background glow to the eight that are visible in a 70-mm binocular. I find it of interest as it only has a few stars in each magnitude band and forms a "mini-coathanger" reminiscent of Brocchi's Cluster (Cr 399).

Ursa Minor: Asterism: The *Engagement Ring* (70 mm)

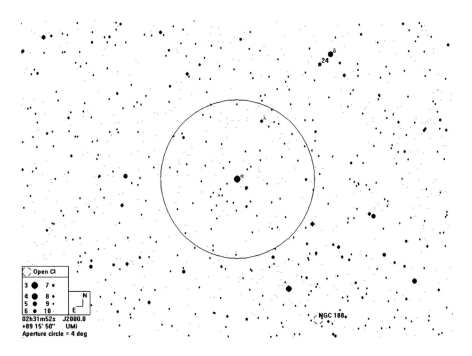

The ***Engagement Ring*** is an asterism that includes ***Polaris*** (α **UMi**).

This is a pretty circlet of mostly 9th magnitude stars that includes the 2nd magnitude ***Polaris*** as the "diamond" in the ring. It is approximately ¾° in diameter and its center is very slightly E of S of *Polaris*, making it useful in locating the North Celestial Pole, which is ¾° N of *Polaris*.

Taurus: Open Cluster: M45 (the *Pleiades*) (50 mm)

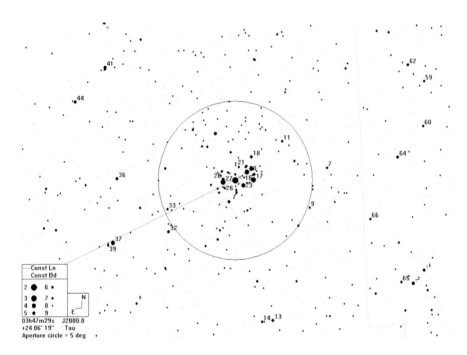

M45 is possibly the most stunning binocular object, one which I never fail to return to each autumn. Putting binoculars onto it is akin to opening a box of diamonds as more of its blue-white members are revealed. Even in 10×50 binoculars it is easy to lose count of them, with several tens of stars being easily visible. In a very dark sky, good quality binoculars will give hints of the nebulosity surrounding *Merope* if you use averted vision. Larger binoculars will show more of the 300 or so stars that comprise it.

Camelopardalis: Asterism: *Kemble's Cascade* (70 mm)

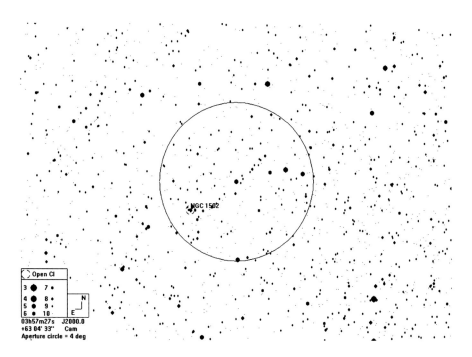

Kemble's Cascade lies in a region of sky that is sparse of bright stars. If you are confident of identifying the 4th magnitude α **Camelopardalis** in your skies, simply find the 5th magnitude star 4° to the SW and then continue the same distance to the SW. If α **Cam** is not visible or identifiable, begin at α **Persei** (*Mirfak*) and scan 14° to the NNE.

This beautiful chain of stars, named for the late Canadian amateur astronomer, Fr Lucien Kemble, is one of the northern sky's finest sights in medium-sized binoculars. It is a ribbon of stars down to 10th magnitude, more than a twenty of which can be visible in 15×70 binoculars, that extends from NW to SE across a 4° field, with a brighter (5th magnitude) star near the middle and the small open cluster NGC 1502 at the SE, which is the "pool" into which the "cascade" appears to "fall."

Appendix 1

Double Stars for Indicating Resolution

The following list of double stars can be used, in addition to those given in the observing lists, as an aid for comparing binocular performance and indicating resolution.

Separation (arcsec)	Star
31	ι Can
20	24 Com
15	ζ UMa
14.4	94 Aqr
11	ε Equ
10	γ Del
7.5	γ Ari
6.5	54 Leo
6	ζ Can

Appendix 1

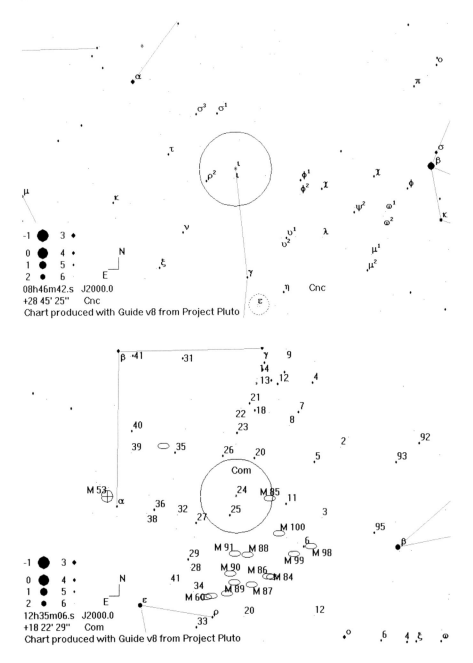

Double Stars for Indicating Resolution

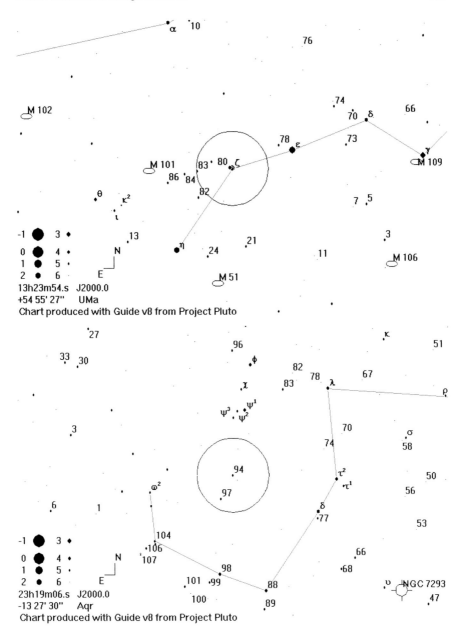

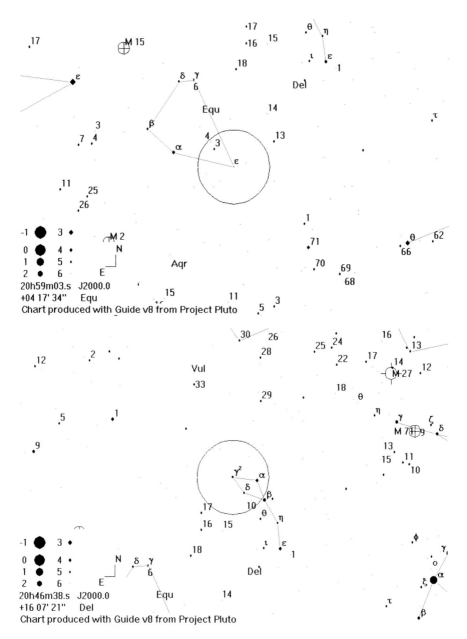

Double Stars for Indicating Resolution

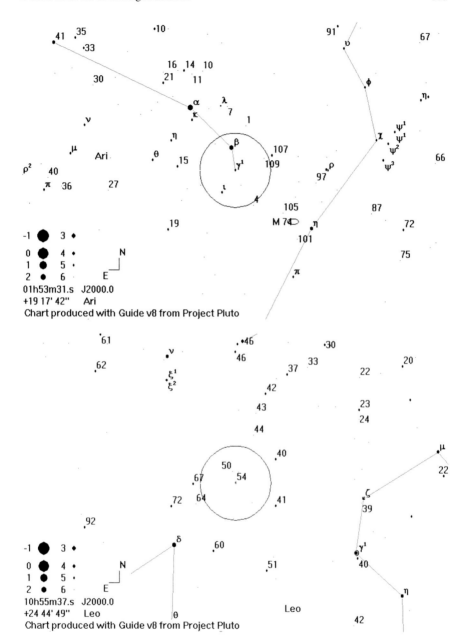

Appendix 1

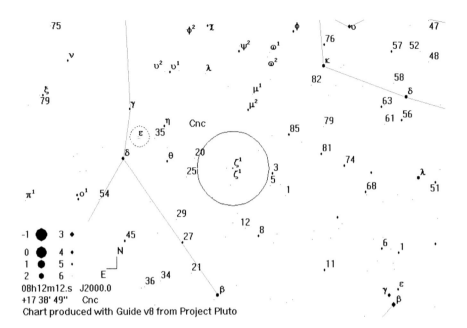

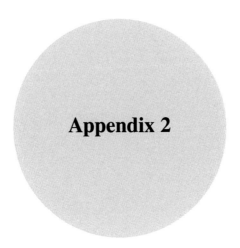

Appendix 2

Limiting Magnitude

Limiting magnitude may be determined by finding the magnitude of the faintest star observable or by counting the stars in a known region of sky. In addition to the optical quality of the instrument, the limiting magnitude will depend upon sky conditions, the experience of the observer, and the altitude of the objects being observed.

M45: The Pleiades

The region to be counted is that bounded by (but not including) Alcyone, Maia, Electra, and Merope.

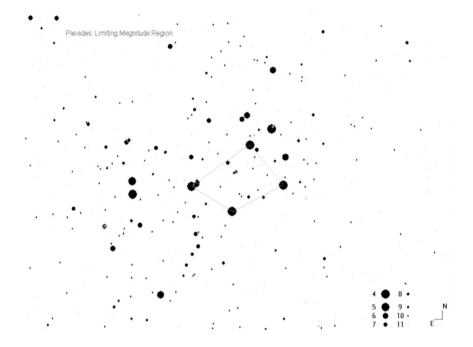

Number of stars	Limiting magnitude
6	9.0
7	9.5
9	10.0
12	10.5
15	11.0
18	11.5
22	12.0
25	12.5
31	13.0

IC2602 (the "Southern Pleiades")

The region to be counted is a triangle bounded (but not including) θ *Carinae* and two 5th magnitude stars.

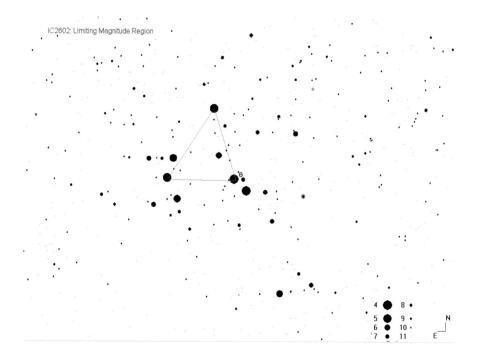

Number of stars	Limiting magnitude
2	9.5
4	10.0
7	10.5
12	11.0
13	11.5
17	12.0
28	12.5
36	13.0

Delphinus

The region to be counted is the "kite" bounded by (but not including) α, β, γ, and δ *Delphini*. Suitable for smaller binoculars.

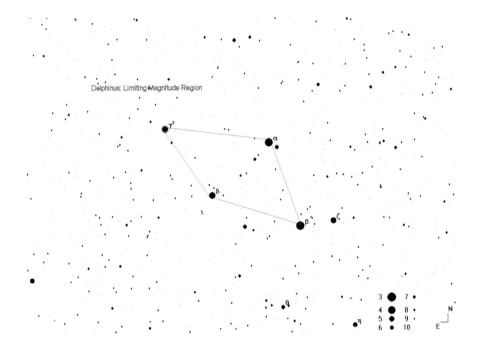

Number of stars	Limiting magnitude
6	8.5
9	9.0
14	9.5
24	10.0
38	10.5

M34

The region to be counted is that bounded by, and including, the parallelogram of 8th magnitude stars. There are a number of optical doubles in this region, so users of larger binoculars (100 mm and above) will get more reliable results at higher magnifications.

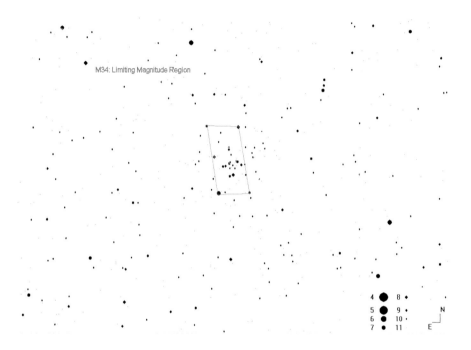

Number of stars	Limiting magnitude
6	8.5
10	9.0
12	9.5
16	10.0
18	10.5
26	11.0
33	11.5
40	12.0
50	12.5
61	13.0

Appendix 3

True Field of View

The charts and tables in this section can be used to help establish the true field of view of binoculars. It is arranged in alphabetical order by the constellation at the center of the field. Each chart shows a $28° \times 20°$ area of sky.

Auriga

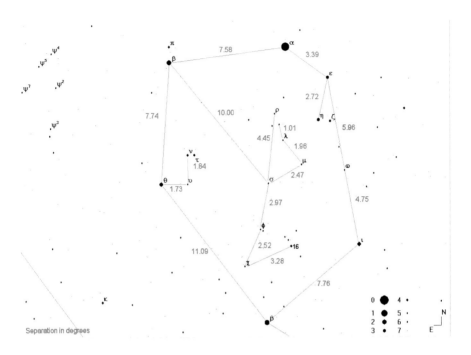

Separation	1st star	2nd star
11.09°	β Tauri	θ Aurigae
10.00°	β Aurigae	σ Aurigae
7.76°	β Tauri	ι Aurigae
7.74°	θ Aurigae	β Aurigae
7.58°	β Aurigae	α Aurigae
5.96°	ε Aurigae	ω Aurigae
4.75°	ω Aurigae	ι Aurigae
4.45°	σ Aurigae	ρ Aurigae
3.39°	α Aurigae	ε Aurigae
3.28°	χ Aurigae	16 Aurigae
2.97°	φ Aurigae	σ Aurigae
2.72°	ε Aurigae	η Aurigae
2.52°	χ Aurigae	φ Aurigae
2.47°	σ Aurigae	μ Aurigae
1.96°	μ Aurigae	λ Aurigae
1.84°	υ Aurigae	ν Aurigae
1.73°	θ Aurigae	υ Aurigae
1.01°	λ Aurigae	BSC 1738

Crux, Centaurus, Musca

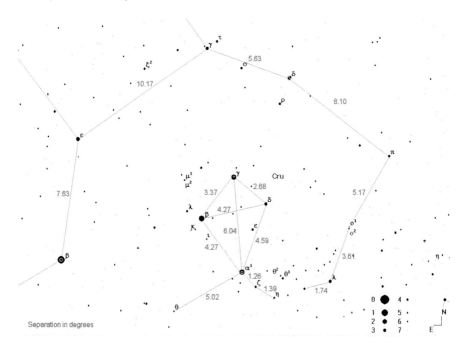

Separation	1st star	2nd star
10.17°	ε Centauri	γ Centauri
8.10°	δ Centauri	π Centauri
7.63°	β Centauri	ε Centauri
6.04°	α Crucis	γ Crucis
5.63°	γ Centauri	δ Centauri
5.17°	π Centauri	o¹ Centauri
5.02°	α Crucis	θ Muscae
4.59°	α Crucis	δ Crucis
4.27°	β Crucis	δ Crucis
4.27°	β Crucis	α Crucis
3.61°	λ Centauri	o¹ Centauri
3.37°	β Crucis	γ Crucis
2.68°	γ Crucis	δ Crucis
1.74°	λ Centauri	BSC 4537
1.39°	η Crucis	ζ Crucis
1.26°	α Crucis	ζ Crucis

Lyra, Cygnus, Hercules

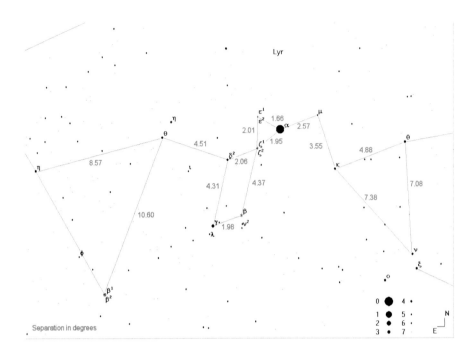

Separation	1st star	2nd star
10.60°	θ Lyrae	β¹ Cygni
8.57°	θ Lyrae	η Cygni
7.38°	κ Lyrae	ν Herculis
7.08°	η Herculis	θ Herculis
4.88°	θ Herculis	κ Lyrae
4.51°	θ Lyrae	δ² Lyrae
4.37°	β Lyrae	ζ¹ Lyrae
4.31°	γ Lyrae	δ² Lyrae
3.55°	κ Lyrae	μ Lyrae
2.57°	α Lyrae	μ Lyrae
2.06°	δ² Lyrae	ζ¹ Lyrae
2.01°	ε² Cygni	ζ¹ Lyrae
1.98°	γ Lyrae	β Lyrae
1.95°	α Lyrae	ζ¹ Lyrae
1.66°	α Lyrae	ε² Cygni

Reticulum, Dorado, and Horologium

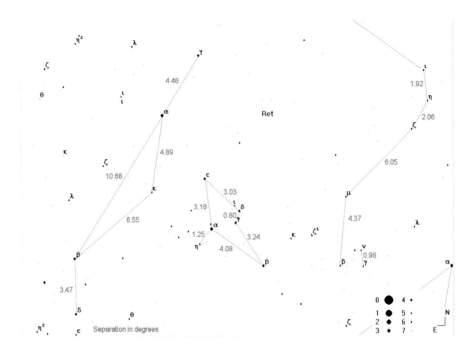

Separation	1st star	2nd star
10.68°	β Doradus	α Doradus
6.55°	β Doradus	κ Doradus
6.05°	ζ Horologii	μ Horologii
4.89°	α Doradus	κ Doradus
4.46°	α Doradus	γ Doradus
4.37°	μ Horologii	β Horologii
4.08°	α Reticuli	β Reticuli
3.47°	β Doradus	δ Doradus
3.24°	β Reticuli	γ Reticuli
3.18°	α Reticuli	ε Reticuli
3.03°	δ Reticuli	ε Reticuli
2.06°	ζ Horologii	η Horologii
1.92°	η Horologii	ι Horologii
1.25°	α Reticuli	η Reticuli
0.98°	γ Horologii	ν Horologii
0.80°	δ Reticuli	γ Reticuli

Sagittarius

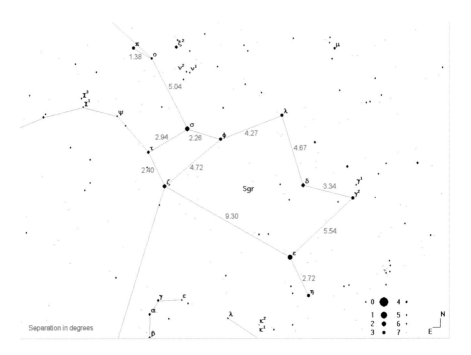

Separation	1st star	2nd star
9.30°	ζ Sagittarii	ε Sagittarii
5.54°	ε Sagittarii	γ² Sagittarii
5.04°	σ Sagittarii	o Sagittarii
4.72°	ζ Sagittarii	φ Sagittarii
4.67°	λ Sagittarii	δ Sagittarii
4.27°	λ Sagittarii	φ Sagittarii
3.34°	δ Sagittarii	γ² Sagittarii
2.94°	σ Sagittarii	τ Sagittarii
2.72°	ε Sagittarii	η Sagittarii
2.40°	τ Sagittarii	ζ Sagittarii
2.26°	σ Sagittarii	φ Sagittarii
1.38°	o Sagittarii	π Sagittarii

Ursa Major

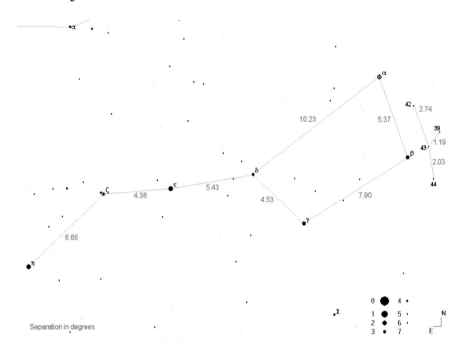

Separation	1st star	2nd star
10.23°	α Ursae Majoris	δ Ursae Majoris
7.90°	β Ursae Majoris	γ Ursae Majoris
6.68°	η Ursae Majoris	ζ Ursae Majoris
5.43°	ε Ursae Majoris	δ Ursae Majoris
5.37°	α Ursae Majoris	β Ursae Majoris
4.53°	γ Ursae Majoris	δ Ursae Majoris
4.36°	ζ Ursae Majoris	ε Ursae Majoris
2.74°	42 Ursae Majoris	43 Ursae Majoris
2.03°	43 Ursae Majoris	44 Ursae Majoris
1.19°	43 Ursae Majoris	39 Ursae Majoris

Ursa Minor

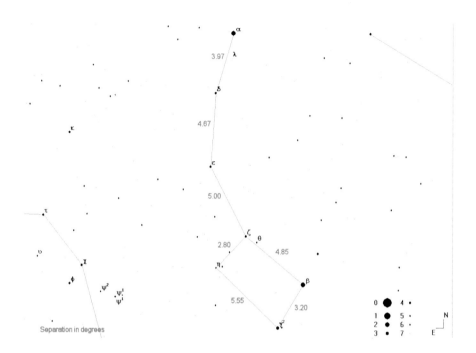

Separation	1st star	2nd star
5.55°	η Ursae Minoris	γ² Ursae Minoris
5.00°	ζ Ursae Minoris	ε Ursae Minoris
4.85°	ζ Ursae Minoris	β Ursae Minoris
4.67°	δ Ursae Minoris	ε Ursae Minoris
3.97°	α Ursae Minoris	δ Ursae Minoris
3.20°	β Ursae Minoris	γ² Ursae Minoris
2.80°	ζ Ursae Minoris	φ Ursae Minoris

Appendix 4

Useful Addresses

Manufacturers

Canon Inc
http://canon.com/

Carl Zeiss GmbH
http://www.zeiss.de/de/bino/home_e.nsf

Fujinon Inc
http://fujinon.com/

Kunming United Optics Corporation
http://www.united-optics.com/

Miyauchi Optical
http://www.miyauchi-opt.co.jp/

Nikon Corporation
http://www.nikon.com/

Starchair
http://www.starchair.com/

Universal Astronomics
http://www.universalastronomics.com/

Software

2Sky: http://open2sky.sourceforge.net/
Planetarium: http://www.aho.ch/pilotplanets/
PleiadAtlas: http://www.astronomycorner.net/PleiadAtlas/
SkySafari: http://www.southernstars.com/
LunaSolCal: http://www.vvse.com/products/en/lunasolcal.html
SkEye: http://lavadip.com/skeye/
AstroPanel: http://astrotips.com/software/astro-panel
Sky Harbinger: http://www.sibimon.net/node/7

Suppliers

Germany

Teleskop-Service Ransburg
Keferloher Marktstraße 19 c
85640 Putzbrunn/Solalinden
Phone: +49 (0)89-1892870
Fax: +49 (0)89-18928710
E-Mail: info@teleskop-service.de
Web: http://www.teleskop-service.de/

UK

Monk Optics Ltd,
Wye Valley Observatory,
The Old School,
Brockweir,
Chepstow,
NP16 7NW
Tel: 01291 689858
Fax: 01291 689834
Web: http://www.monkoptics.co.uk

Optical Vision Ltd
Unit 2b, Woolpit Business Park,
Woolpit,
Bury St. Edmunds,
Suffolk IP30 9UP,
England
Tel: +44 (0)1359 244 200
Fax: +44 (0)1359 244 255
Email: info@opticalvision.co.uk
Web: http://www.opticalvision.co.uk/

Opticron
Unit 21, Titan Court,
Laporte Way,
Luton,
Bedfordshire, LU4 8EF,
UK
Tel: 01582 726522
Fax : 01582 723559
Email: info@opticron.co.uk
Web: http://www.opticron.co.uk/

Strathspey Binoculars
Robertland Villa Railway Terrace
Aviemore
PH22 1SA
Scotland, UK
Tel +44 (0)1479 812549
Email: john@unixnerd.demon.co.uk
Web: http://www.strathspey.co.uk/

USA

The Binoscope Company
3 Wyman Court
Coram, NY 11727
Phone: 631-473-5349
Fax: 631-331-8891
Email: astrojoe@optonline.net
Web: http://www.binoscope.com/

Hutech Corporation
25691 Atlantic Ocean Dr., Unit B-11
Lake Forest, CA 92630
Phone: 877-BUY-BORG (toll free)
Fax: 949-859-5512 (fax)
Email: info@hutech.com
Web: http://www.hutech.com/

Garrett Optical LLC
1611S. Utica Ave. #323
Tulsa, OK 74104
Phone: 888-629-2011
Fax: 888.-48-1662
Email: garrett@garrettoptical.com
Web: http://garrettoptical.com

Jim's Mobile Incorporated
8550 West 14th Avenue
Lakewood, CO 80215
U.S.A.
Phone: 303-233-5353
Fax: 303-233-5359
Email: info@jmitelescopes.com

Oberwerk Corporation
75-C Harbert Dr.
Beavercreek, OH 45440
Phone: 937-426-8892
Email: info@oberwerk.com

Orion Telescopes and Binoculars
89 Hangar Way
Watsonville, CA 95076
Phone: 831-763-7000
Email: support@telescope.com
Web: http://www.telescope.com/

Universal Astronomics
6 River Ct.
Webster, MA 01570
Phone: 508-943-5105
Fax: 707-371-0777
Email: Larry@UniversalAstronomics.com
Web: http://www.universalastronomics.com/

Binocular Repair

UK

Action Optics
16 Butts Ash Gardens,
Hythe,
Southampton.
SO45 3BL.
Telephone or Fax: 023 8084 2801
Mobile: 079 77 88 1482.
Email: richard@actionoptics.co.uk
Web: http://www.actionoptics.co.uk

OptRep
16 Wheatfield Road
Selsey
West Sussex
PO20 0NY
Phone: 01243 601 365
Fax: 01243 601 365
Email: info@opticalrepairs.com
Web: http://www.binocularrepairs.co.uk

USA

Captains Nautical Supplies, Inc.
2500 15th Avenue West
Seattle, WA 98119
Phone: 206-283-7242, Toll-free in the USA and Canada 800-448-2278
Fax: 206-281-4921
E-Mail: sales@captainsnautical.com

Suddarth Optical Repair
205 West May St.
Henryetta, OK 74437
Phone: 918-650-9087
E-Mail: binofixer@aol.com

Appendix 5

Internet Resources

AstroClassifieds
http://www.astroclassifieds.co.uk/

Astromart Astronomy Classifieds
http://www.astromart.com/

Astronomy Centre
http://www.astronomycentre.org.uk/

Binocular Astronomy Resource Page
http://www.uvaa.org/binocularresources.htm

Binocular Sky
http://binocularsky.com/

Canada-wide Astronomy Buy & Sell
http://www.astrobuysell.com/

Cloudy Nights Binoculars Forum
http://www.cloudynights.com/ubbthreads/postlist.php?Cat=0&Board=binoculars

Stargazers Lounge Binocular Forums
http://stargazerslounge.com/forum/80-observing-with-binoculars/
http://stargazerslounge.com/forum/133-discussions-binoculars/

Telescopes and Astronomy Supplies in Australia
http://www.quasarastronomy.com.au/shops.htm

UK Astronomy Buy & Sell
http://www.astrobuysell.com/uk/

Appendix 6

Binocular Designations

There is a potentially bewildering array of letters that manufacturers use to give further information about binoculars in addition to the magnification and aperture. There is not an industry-wide standard of these, but here are most of the ones in recent and current use:

- **A:** Armoured, usually with rubber (see **GA**, below).
- **AG:** Silver coating on reflective surfaces of roof prisms (from *argentum*, the Latin for *silver*, whose chemical symbol is similar: *Ag*)
- **B:** Depends on context and/or source of binoculars.
  - (a) Usual American, Chinese, and Japanese usage. A Porro-prism binocular with each optical tube of one-piece construction (*Bausch & Lomb* or *American* style)
  - (b) Usual European usage: Long eye relief, suitable for spectacle-wearers (from *brille*, the German for *spectacles*).
- **C:** Depends on context.
  - (a) Compact binocular (usually a small roof-prism binocular).
  - (b) Coated optics.
- **CF:** Center focus. Usually combined with another letter, e.g., BCF: *Bausch & Lomb* style center focus.
- **D:** Roof prism binocular (from *dach*, the German for *roof*).
- **F:** Flat-field technology
- **FC:** Fully coated optics.
- **FL:** Fluorite lenses.

FMC:	Fully multicoated optics.
GA:	Rubber armored (from *gummi*, the German for *rubber*)
H:	H-body roof prism binocular.
IF:	individual focusing eyepieces. Usually combined with another letter, e.g. ZIF: *Zeiss* style individual focus.
IS.	Image stabilized.
MC:	Multicoated optics.
MCF:	Mini center-focus ("delta" Porro-prism binoculars, with the objectives closer together than the eyepieces)
N:	Nitrogen filled
P or **PC**:	Phase-corrected prism coatings (on roof-prisms).
P*:	Proprietary (to *Zeiss*) phase coatings
SMC:	Fully multicoated optics.
T*:	Proprietary (to *Zeiss*) anti-reflective coatings
W, **WA**, or **WW**:	Wide angle.
WP:	Waterproof.
Z:	Porro-prism binocular with each optical tube of two-piece construction, with the objective tube screwing into the prism housing (*Zeiss* or *European* style).

T*, FMT, RC, SL, SLC

Appendix 7

Glossary of Terms

Abaxial Rays: Rays that are distant from the optical axis

Abbé Prism: A roof prism.

Aberration: An optical effect which degrades an image.

Accommodation: The ability of the eyes to focus on both near and distant objects. The normal range is from about 120 mm (4½ in.) to infinity.

Achromatic: Literally "no colour". A lens combination in which *chromatic aberration* is corrected by bringing two colors to the same focus.

Absolute Magnitude: The *apparent magnitude* that an object would possess it if were placed at a distance of 10 *parsecs* from the observer. In this way, absolute magnitude provides a direct comparison of the brightness of stars.

Airy Disc: The bright central part of the image of a star. It is surrounded by diffraction rings and its size is determined by the aperture of the telescope or binocular. About 85 % of the light from the star should fall into the Airy disc. You will not see the disc and rings unless you have a binocular telescope or binoculars that operate at unusually high magnifications.

Altazimuth: A mounting in which the axes of rotation are horizontal and vertical. An altazimuth mount requires motion of both axes to follow an astronomical object, but is simpler to make than an *equatorial mount* and can, in some forms, be held together by gravity.

Altitude: The angle of a body above or below the plane of the horizon—negative altitudes are below the horizon.

Albedo: The proportion of incident light which a body reflects in all directions. The albedo of Earth is 0.36, that of the Moon is 0.07 and that of Uranus is 0.93. The true albedo may vary over the surface of the object, so, for practical purposes, the mean albedo is used.

Amici Prism: A right-angled prism whose hypotenuse face has been formed into a roof. It is used to erect images.

Anastigmat: An optical system that is corrected for *astigmatism* in at least one off-axis zone and for which it has tolerable correction for the intended purpose over the rest of the field.

Angle of Incidence: The angle between the *normal* to an optical surface and the incident ray.

Angle of Reflection: The angle formed between the *normal* to an optical surface and the reflected ray.

Angle of Refraction The angle formed between the *normal* to an optical surface and the refracted ray.

Aperture: The diameter of the largest bundle of light that can *enter* an optical system. It is usually the diameter of objective lens or primary mirror.

Aperture Stop: A physical aperture that restricts the size of the bundle of light passing *through* an optical system.

Aphelion: The position in a *heliocentric* orbit at which the orbiting object is at its greatest distance from the Sun.

Apoapsis: The position in an orbit at which the orbiting object is at its greatest distance from the object about which it is orbiting.

Apochromatic: A lens combination in which *chromatic aberration* is corrected by bringing three colors to the same focus. The term is used by some manufacturers to describe *achromatic* doublets whose false color is approximately equivalent to that of an apochromatic triplet lens.

Apogee: The position in a *geocentric* orbit at which the orbiting object is at its greatest distance from Earth.

Apparent Field (of View): The angular size of the entire image (see *True Field (of View)*).

Apparent Magnitude: The brightness of a body, as it appears to the observer, measured on a standard *magnitude* scale. It is a function of the *luminosity* and distance of the object and the transparency of the medium through which it is observed.

Arc minute: One sixtieth of a degree of arc.

Arc second: The second division of a degree of arc. One sixtieth of an arc minute (1/3,600th of a degree).

Astigmatism: An optical *aberration* resulting from unequal magnification across different diameters.

Axial Rays: Rays that originate from a distant object on the optical axis.

Azimuth: The angular distance around the horizon, usually measured from north (although it is sometimes measured from south), of the *great circle* passing through the object.

Back Focus: The distance between the exit aperture of an optical tube and the position of the image plane. Binoviewers in particular require a significant amount of back focus.

Baffle: An opaque barrier that is positioned so as to reduce or eliminate the effects of stray light in an optical system. In binoculars these often take the form of machined ridges or screw threads in the objective tubes.

BaK4: Barium Crown glass. This is a glass of high optical density that is used for the prisms of good quality binoculars.

Barlow Lens: A diverging lens which has the effect of increasing (usually doubling) the effective focal length of the telescope.

Binoviewer: A device that splits equally the single light cone from a telescope into two light cones in order that both the observer's eyes may be used for observation.

BK7: Borosilicate glass. Cheaper and less dense than **BaK4**, it is commonly found in cheaper binocular prisms.

Blind Spot: The position where the optic nerve enters the retina. It can become noticeable when using optical instruments, but its effect is ameliorated by the use of both eyes.

Binocular Vision: See **Stereopsis**.

Catadioptric: A telescope whose optics, not including the eyepiece, consists of both lenses and mirrors. The most common examples of these are the Schmidt-Cassegrain telescopes, whose "lens" is an aspheric corrector plate, and the Maksutov-Cassegrain telescopes, whose "lens" is a deeply curved meniscus.

Cells: The part of an optical instrument that holds the lenses or mirrors.

Chromatic Aberration: An *aberration* of refractive optical systems in which light is dispersed into its component colors, resulting in false color in the image. There are two distinct manifestations of it:
  (i) *Longitudinal Chromatic Aberration*: Light of different wavelengths is brought to different foci.
  (ii) *Lateral Chromatic Aberration*: Light of different wavelengths forms images of different sizes.

Collimation: 1. The act of bringing of the optical components of a telescope into correct alignment with each other.
  2. The act of making the optical axes of the optical tubes of the binocular parallel to each other and, where appropriate, to the central hinge (see also ***Conditional Alignment***).

Coma: (i) The matter surrounding the nucleus of a comet—it results from the evaporation of the nucleus.
  (ii) An optical *aberration* in which stellar images are fan shaped, similar to comets.

Comes (pl comites): A member of a multiple star system.

Concave: Curving inwards at the center.

Convex: Curving outwards at the center.

Conditional Alignment: An incomplete *collimation* of a binocular in which the optical axes are parallel to each other but not to the central hinge. The optical axes are therefore only parallel at a specific ***inter pupillary distance***.

Culmination: An object is at culmination when it reaches the observer's meridian. It is then at its greatest altitude.
Declination: The angle of an object above or below the *celestial equator*. It is part of the system of equatorial coordinates.
Depth of Field: The range of distances from the objective lens for which objects appear to be in focus when the binocular (or other optical system) is focused on an object within that range.
Dialyte: A doublet lens in which the inner surfaces have different curvatures and which cannot therefore be cemented.
Diffraction Limited: A measure of optical quality in which the performance is limited only by the size of the theoretical diffracted image of a star for a telescope of that aperture.
Distortion: An aberration in which the periphery of the field undergoes a different magnification to the center of the field. There are two major types:
Barrel distortion: **The center is magnified more than the periphery.**
Pincushion distortion: **The periphery is magnified more than the center.**
Dobsonian Named after John Dobson, who originated the design. An *altazimuth mount* constructed usually of plywood or MDF suited to home construction. Also refers to a telescope or binocular telescope so mounted.
Elongation: The angular distance between the Sun and any other solar system body or between a satellite and its parent planet.
Equatorial Mount: A mounting in which one of two mutually perpendicular axes is aligned with Earth's axis of rotation, thus permitting an object to be tracked by rotating this axis so that it counteracts Earth's rotation.
Exit Pupil: The position of the image of the aperture formed by the eyepiece. It is the smallest disc through which all the collected light passes and is therefore the best position for the eye's pupil. Also known as an *eye ring* or a *Ramsden disc*.
Eye Relief: The distance from the eye lens of the eyepiece to the *exit pupil*. Spectacle wearers require sufficient eye relief to enable them to place the eye at the *exit pupil*.
Eyepiece: The lens combination that is closest to the eye.
Eye Ring: An alternative name for the *exit pupil*.
Field of View: The maximum angle of view through an optical instrument.
f-Number, f-Ratio: The ratio of the focal length to the aperture.
Focal Plane: The plane (usually this is actually the surface of a sphere of large radius) where the image is formed by the main optics of the telescope. The eyepiece examines this image.
Focuser: The part of the telescope which varies the optical distance between the objective lens or primary mirror and the eyepiece. This is usually achieved by moving the eyepiece in a drawtube, but in some catadioptric telescopes, it is the primary mirror which is moved.
Fork Mount: A mount where the telescope swings in declination or in altitude between two arms. It is suited only to short telescope tubes, such as Cassegrains, and variations thereof. It requires a *wedge* to be used equatorially.

Galilean Moons: The four Jovian moons first observed by Galileo (Io, Europa, Ganymede, and Callisto). They are observable with small binoculars.

German Equatorial Mount (GEM): A common *equatorial mount* for small- and medium-sized amateur telescopes, suited to both long and short telescope tubes. The telescope tube is connected to the counterweighted declination axis, which rotates in a housing which keeps it orthogonal to the polar axis. Tracking an object across the meridian requires that the telescope be moved from one side of the mount to the other, which in turn requires that both axes are rotated through 180°, thus reversing the orientation of the image. This is not a problem for visual observation but is a limitation for astrophotography.

Granulation: The "grains of rice" appearance of the Sun's surface, which results from convection cells within the Sun.

Great Circle: A circle formed on the surface of a sphere which is formed by the intersection of a plane which passes through the center of a sphere. A great circle path is the shortest distance between two points on a spherical surface.

Inferior Conjunction: The *conjunction* of Mercury or Venus when they lie between Earth and the Sun.

Inferior Planets: Planets (i.e., Mercury and Venus) whose orbits lie inside Earth's orbit.

Image Plane: A plane, perpendicular to the optical axis, where the image is formed.

Infinity: The distance at which rays from an object are indistinguishable from parallel. It has the symbol?

Inter Pupillary Distance: The distance between the pupils of the eye when the observer is viewing a distant object or between the exit pupils of a binocular. The latter needs to be adjustable so as to match the former.

Inverted: Upside down.

Light Bucket: Slang term for a telescope of large aperture.

Light-Year: The distance travelled by light in 1 year: 9.4607×10^{12} km or 63,240 AU or 0.3066 *parsecs*.

Limb: The edge of the disc of a celestial body.

Luminosity: The amount of energy radiated into space per second by a star. The bolometric luminosity is the total amount of radiation at all frequencies; sometimes luminosity is given for a specific band of frequencies (e.g., the visual band).

Magnitude: The brightness of a celestial body on a numerical scale. See also *absolute magnitude*, *apparent magnitude, bolometric magnitude,* and *integrated magnitude*.

Meridian: The *great circle* passing through the celestial poles and the observer's *zenith*.

Minor Planets: Another term for asteroids.

Night Glass: A binocular (or telescope) with an **exit pupil** of 7 mm or more.

Normal: Perpendicular to an optical surface.

Normally: Impinging perpendicularly on an optical surface.

Occultation: An alignment of two bodies with the observer such that the nearer body prevents the light from the further body from reaching the observer. The nearer body is said to occult the further body. A solar *eclipse* is an example of an occultation.

Off-Axis: At an angle to the optical axis.

Opposition: The position of a planet such that Earth lies between the planet and the Sun. Planets at opposition are closest to Earth at opposition and thus opposition offers the best opportunity for observation.

Optical Axis: The "line of optical centers" of the elements of an optical system. It is the line of the principle axes of these optical elements and, when they are curved, it is the line passing through their centers of curvatures.

OTA: Abbreviation for *optical tube assembly*. It is normally considered to consist of the tube itself, the focuser and the optical train from the objective lens (refractor), primary mirror (reflector), or corrector plate (catadioptrics) up to, but not including, the eyepiece.

Paraxial Rays: Rays that are close to and parallel to the optical axis.

Parfocal Eyepieces: Eyepieces sharing the same focal plane. They can be interchanged without requiring refocusing.

Parsec: The distance at which a star would have a parallax of 1 arcsec (3.2616 *light-years*, 206,265 *astronomical units*, 30.857×10^{12} km).

Pechan Prism: A prism, consisting of two air-spaced elements, that will revert an image without inverting it.

Periapsis: The position in an orbit at which the orbiting object is at its least distance from the object about which it is orbiting.

Periastron: The position in an orbit about a star at which the orbiting object is at its least distance from the star.

Perigee: The position in a *geocentric* orbit at which the orbiting object is at its least distance from Earth.

Perihelion: The position in a *heliocentric* orbit at which the orbiting object is at its least distance from the Sun.

Phase: The percentage illumination, from the observer's perspective, of an object (normally planet or Moon).

Phase Coating: A coating used on **roof prism** binoculars to increase contrast by correcting a differential phase shift.

Planisphere: The projection of a sphere (or part thereof) onto a plane. It commonly refers to a simple device which consists of a pair of concentric discs, one of which has part of the *celestial sphere* projected onto it and the other of which has a window representing the horizon. Scales about the perimeters of the disc allow it to be set to show the sky at specific times and dates, enabling its use as a simple and convenient aid to location of objects.

Porro-Prism: An isosceles right-angled prism that reflects the light, by total internal reflection, off both shorter faces, giving a combined angle of reflection of 180°.

Prism: A transparent body with two or more optically flat surfaces that are inclined to each other. Light is either refracted or reflected at these surfaces.

Proper Motion: The apparent motion of a star with respect to its surroundings.

Quadrature: The position of a body (Moon or planet) such that the Sun-body-Earth angle is 90°. The *phase* of the body will be 50 %.

Rayleigh Criterion (Rayleigh Limit): Lord Rayleigh, a nineteenth-century physicist, showed that a telescope optic would be indistinguishable from a theoretical perfect optic if the light deviated from the ideal condition by no more than one quarter of its wavelength.

Reflector: A telescope whose optics, apart from the eyepiece, consist of mirrors.

Refractive Index: A measure of the relationship between the angles of incident and refracted rays.

Refractor: A telescope whose optics consist entirely of lenses.

Resolution: A measure of the degree of detail visible in an image. It is normally measured in arc seconds.

Reticle: A system of engravings in a transparent disc, or of wires or hairs, placed at the focal plane of the eyepiece so as to superimpose a grid or other pattern over the field of view.

Reverted: Laterally reversed.

Rhomboidal Prism: A reflecting prism which has two parallel reflecting surfaces and two parallel transmitting surfaces. It is used to offset an image without changing its orientation. Pairs of rhomboidal prisms are used to adjust the *inter pupillary distance* in b*inoviewers* and some binocular telescopes.

Right Ascension (RA): The angle, measured eastward on the *celestial equator*, between the *First Point of Aries* and the *hour circle* through the object.

Roof Prism: A prism in which one face has been formed into a "roof" with a right-angled apex.

Scintillation: The twinkling of stars, resulting from atmospheric disturbance.

Spherical Aberration: An optical *aberration* in which light from different parts of a mirror or lens is brought to different foci.

Stereopsis, Stereoscopic Vision: Three-dimensional vision that results from the spacing of the eyes, each eye seeing the object from a slightly different angle.

Superior Conjunction: The *conjunction* of Venus and Mercury when they are more distant than the Sun.

Superior Planets Those planets whose orbits lie outside Earth's orbit.

Terminator: The boundary of the illuminated part of the disc of a planet or moon.

Transit: (i) The passage of Mercury or Venus across the disc of the Sun
  (ii) The passage of a planet's moon across the disc of the parent planet
  (iii) The passage of a planetary feature (such as Jupiter's Great Red Spot) across the *central meridian* of the planet
  (iv) The passage of an object across the observer's *meridian* (see *culmination*)

True Field (of View): The angular size of the entire object. It is the angle of the cone of rays at the aperture that is transmitted as a usable image.

Umbra: (i) The shadow that results when a bright object is completely occulted. A total *eclipse* of the Sun occurs when the observer is in the Moon's umbra.
  (ii) The dark inner region of a sunspot.

Vignetting: The loss of light, usually around the periphery of an image, as a consequence of an incomplete bundle of rays passing through the optical system.

Visual Axis: A line from the object, through the node of the eye's lens, to the fovea.

Wedge: The part that fits between the tripod or pillar and the fork of a fork-mounted telescope, which enables the fork to be equatorially aligned.

Zenith: The point on the *meridian* directly above an observer.

Index

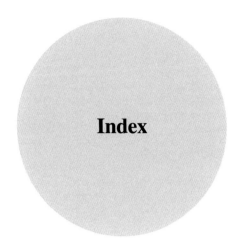

A
Abbé erecting system, 14–16
Abbe-König roof prism, 21
Abbé number, 16, 17
Aberrations, 6–8, 10, 11, 21, 27, 29–35, 43, 49, 57, 63–65, 68, 70, 71
Achromatic, 7, 11, 29, 70
Adler, Alan, 45
ADS 6915, 232
ADS 6921, 232
Albireo, 172, 177, 185, 334
Alcock, George, viii
Aldebaran, 197, 198, 380
Algol, 384, 386
Alpha Persei Moving Cluster, 173, 176, 188, 385
Anaglyph, 73, 74
Android, 136, 137
56 Andromedae, 172
Angled eyepieces, 12, 47, 55, 103, 107, 117, 129, 141
Aphrodite, 259
Apochromatic, 11, 29
Argo Navis, 125
14 Arietis, 172, 176, 182, 378
Aristotle, 219
Asterisms, 169, 203, 248, 250, 307, 308, 314, 327, 330, 362, 370, 374, 387, 388
Astigmatism, 29, 31–33, 40, 70–71
Astro Index, 45–46

Averted vision, 144, 194, 208, 212–215, 230, 235, 249, 251, 252, 255, 267–272, 274, 276, 292–296, 308, 311, 312, 315, 317, 328, 337, 338, 350, 359, 361, 367, 379, 388

B
Bahtinov mask, 74, 75
BaK4, 15–18, 43, 44
Barnard 142, 170, 181, 329
Barnard 143, 170, 181, 329
Barnard 353, 170, 177, 185, 336
Barnard's Star, 175, 178, 187, 298
Bausch & Lomb body, 44, 83, 87, 92
Binoback, 122, 123
Binocular advantage, 5–6, 125
Binocular summation, 5, 58
Binocular telescopes, 31, 48, 60, 119–127, 135, 137
Binoscope, xi, 119, 122, 123
Binoviewer, 17, 21, 23, 58–60, 132
Bishop, Roy, 45
BK7, 15–18, 44, 67
50 Boötis, 172, 180, 182, 280
Boots, 143
Borg, 119, 120
Boyd, Florian, xi, 110
Brocchi's Cluster, 169, 177, 192, 330, 386
Burr, Jim, xi

Butler, Norman, xi, 123
Butterfly Cluster, The, 174, 177, 190, 289

C

Carbon Star, 254
Cases, 4, 43, 57, 63, 66–69, 72, 77, 79–81, 89, 100, 103, 107, 125, 129, 130, 136, 138, 139, 176, 270, 295, 306, 315
Centaurus A, 171, 180, 183, 274
Center focus, 36, 38, 53, 54, 61, 65, 68, 97, 144
Cepheid variables, 313, 358
Chart(s), ix, xii, 5, 49, 74, 113, 135–139, 151, 176, 205, 250, 267, 298, 300, 310, 311
Charting software, 135–137
Christmas Tree Cluster, The, 173, 178, 187, 216
Chromatic aberration, 7, 10, 11, 29, 30, 43, 63, 70
Cleaning, 76, 78–83, 88, 106, 111
Cleaning fluid, 81
Clothing, 76, 81, 82, 106, 142
Coathanger, The, 169, 177, 192, 330
Coatings, 10, 20, 21, 24–28, 44, 46, 58, 65, 81, 82
Collimation, xi , 7, 37–40, 58, 60, 61, 63, 72–75, 87, 91–94, 122–123, 125
Collinder39, 173, 176, 188
Collinder65, 173, 176, 188, 207
Collinder70, 173, 176, 188
Collinder399, 177, 192
Collinder463, 173
Collinder, Per, 150
Coma, 11, 29, 31–33, 70, 184, 213, 259–263, 267, 269
Comfort, 68, 117, 141–143
Compass, 135, 136, 138
Condensation, 21, 36, 77–79, 133
Conditional alignment, 39, 91
Cone Nebula, The, 216
Convergence, 40, 72, 94
Crab Nebula, The, 175, 179, 191, 200
Cook, William J (Bill), ixi, 62,93
Critical angle, 16, 17
β Cygni, 172, 334
61 Cygni, 173, 185, 339

D

Dark adaptation, 137, 144
Delta Boötis, 172, 180, 182, 280
Deneb, 336, 338, 340
Desiccants, 79–80
Detection threshold, 125
Dew, 78, 79, 111, 131, 133–135

Dew heaters, 111, 134
Diopter adjustment, 36, 37, 65, 69, 89, 125
Dipvergence, 40, 61
Dismantling, 79, 82–88, 138
Disposability, 64, 83
Distortion, 29, 34, 43, 63, 71, 73
Divergence, 39, 40, 72, 94
30 Doradus, 170, 179, 185, 205
Double Cluster, 173, 176, 188, 382, 383
Double-handed hold, 98, 99
Dreyer, Johan, 150
Dumbell Nebula, The, 177, 192, 331

E

Emission nebulae, 149, 170, 194, 205, 213, 242, 307, 311, 319, 338
Engagement Ring, The, 169, 177, 191, 387
Epsilon Pegasi, 173, 177, 188, 342
Epsilon Sagittae, 189, 332
Erecting mirror system, 123
Erfle eyepiece, 12
Eta Carinae, 170, 177, 183, 238, 242
ET Cluster, The, 173, 179, 183, 370
Exit pupil, 6–7, 16, 18, 39, 44, 49, 50, 53, 57, 60, 67, 68, 71
Eyepiece, 6–9, 11–12, 21, 29, 34, 36, 37, 43, 44, 47, 49, 50, 53–55, 57–61, 65–69, 71–74, 76–81, 84–86, 88–90, 92–94, 96, 97, 100, 103, 107, 108, 111, 113, 117, 119, 122–124, 129–132, 134, 135, 139, 141, 142, 144, 149, 194, 240, 345, 352
Eye relief, 11, 12, 50–53, 67, 131, 132

F

Field curvature, 29, 33–34, 43, 69
Field of View, 48–50, 74, 255
Field tests, 73–75
Filters, 132–133
Finder scopes, 131
Flashlights, 137–138
Floaters, 60
Floyd, Chris, xi
Fluorite, 8, 11, 29
Flying shadows, 49, 50, 71
Focal range, 65, 69
Focal ratio, 7–8, 11, 14, 75, 119
Focus(ing), 9, 21, 36–38, 44, 54, 65, 70–72, 74, 84, 89, 97, 101, 121–123, 125, 138, 144
Focusing mechanisms, 9, 36–37, 53, 57, 122
Footwear, 143
Fork mounts, 109–110
Fungal growth, 76

G

Galaxies, 3, 6, 50, 136, 150, 170–171, 176, 178–181, 184–186, 190–192, 205, 222, 235, 239, 243–246, 252, 253, 255–257, 260–262, 267–271, 273, 274, 276–278, 353–355, 358, 359, 361, 377, 379, 381
Gamma Leporis, 172, 176, 177, 187, 202
Garnet Star, 176, 177, 184, 345, 346
Ghost of Jupiter, The, 175, 179, 186, 240
Globular clusters, 150, 171–172, 177, 186, 258, 263–265, 268, 270, 272, 275, 281, 286, 302–305, 312, 314–318, 324, 325, 333, 341–344, 354, 357, 358, 381
Gloves, 83, 143
Great Orion Nebula, 170, 176, 188, 242

H

Halley, Edmund, 302
Harlow, Keith, xi, 125, 126, 127
Hats, 142, 143
Helix Nebula, The, 175, 181, 350
Herschel, William, 149, 346, 380
Hind's Crimson Star, 175, 177, 187, 201
Hinge clamp, 100–101
Hyades, The, 150, 173, 176, 190, 197, 198
Hyakutake, Yuji, viii
Hydration, 143

I

IC 1396, 174, 177, 184, 345
IC 1805, 374
IC 2157, 214
IC 2391, 174, 177, 192, 231
IC 2602, 174, 177, 183, 241
IC 2944, 174, 179, 183, 251
IC 4665, 175, 178, 187, 297, 299
IC 4725, 174, 181, 189, 313
IC 4756, 174, 177, 190, 320
Image stabilization, 22, 23, 54, 55
Independent focus, 36–37, 39, 54
Interchangeable eyepieces, 6, 12, 47, 57, 113, 119
Internal reflections, 16, 69
Interpupillary distance (IPD), 4, 15, 19, 22, 37, 39, 65–66, 68, 70, 88, 91, 122, 123
iOS, 136, 137

J

Jewel Box, The, 174, 177, 185, 266

K

Kellner eyepiece, 11
Kemble's Cascade, 169, 176, 192, 389
Kidney beaning, 49, 72
Kunming, 12, 26

L

Lagoon Nebula, The, 170, 177, 189, 307, 308
Large Magellanic Cloud, 170, 179, 185, 205
Lasers, 117, 125, 131, 132
La Superba, 175, 177, 185, 254
L-brackets, 55, 67, 100–103, 139
LDN 906, 170, 177, 185, 336
Lean, 12, 92, 93
Leaping Minnow, The, 169, 176, 182, 203
Leavitt, Henrietta, 358
Lens Pen, 81
Lens tissue, 81, 82
Limiting magnitude, 74
Loop Nebula, The, 170, 179, 185, 205
Lubricants, 65, 80, 89

M

M1, 175, 179, 191, 200
M2, 172, 177, 181, 343, 344
M3, 172, 178, 185, 258
M4, 172, 180, 189, 286
M5, 172, 180, 190, 281, 343
M6, 174, 177, 190, 289
M7, 174, 177, 190, 289, 290
M8, 170, 177, 189, 307, 308
M10, 172, 178, 187, 292, 293
M11, 174, 177, 190, 323
M12, 172, 178, 187, 292, 293
M13, 172, 177, 186, 263, 281, 302, 303, 314, 342, 343
M14, 172, 178, 187, 296
M15, 172, 177, 188, 342
M16, 174, 180, 190, 311, 319
M17, 170, 180, 189, 310, 311
M18, 174, 180, 189, 310, 311
M19, 172, 178, 187, 295
M20, 170, 180, 189, 307, 308
M22, 172, 178, 189, 314
M23, 174, 178, 189, 306
M24, 174, 177, 189, 309
M25, 174, 181, 189, 313
M26, 174, 178, 190, 322

M27, 175, 177, 192, 331
M28, 172, 178, 189, 312
M29, 174, 178, 185, 335
M31, 170, 176, 181, 359
M32, 359
M33, 3, 6, 170, 176, 191, 355, 377
M34, 173, 176, 188, 376, 384
M35, 48, 49, 144, 173, 176, 186, 214
M36, 173, 178, 182, 204
M37, 173, 178, 182, 204
M38, 173, 178, 182, 204
M39, 174, 178, 185, 340
M40, 169, 179, 191, 248
M41, 173, 176, 183, 219
M42, 170, 176, 188, 205, 208
M43, 170, 176, 188, 208
M44, 174, 177, 182, 232
M45, 173, 176, 190, 241, 388
M46, 174, 176, 189, 221
M47, 174, 176, 189, 221
M48, 174, 178, 186, 230
M49, 171, 178, 192, 268
M50, 173, 176, 187, 217
M52, 175, 181, 183, 351, 366
M53, 172, 180, 184, 263
M54, 172, 181, 189, 315
M55, 172, 181, 189, 317
M57, 149
M58, 171, 180, 192, 269
M59, 171, 178, 192, 270
M60, 171, 178, 192, 270
M62, 172, 180, 187, 294, 295
M63, 171, 178, 185, 256
M64, 171, 178, 184, 262
M65, 171, 179, 187, 246
M66, 171, 179, 187, 246
M67, 174, 178, 182, 233
M68, 171, 180, 186, 272
M71, 172, 189, 333
M77, 170, 179, 184, 381
M78, 170, 178, 188, 213
M79, 202
M81, 170, 179, 191, 235
M82, 170, 179, 191, 235
M83, 171, 180, 186, 273
M84, 171, 180, 192, 267–269
M86, 171, 180, 192, 267, 269
M87, 171, 178, 192, 269
M88, 171, 180, 184, 269
M89, 171, 178, 192, 269
M90, 171, 180, 192, 269
M91, 171, 180, 184, 269
M92, 172, 180, 186, 303
M93, 174, 178, 189, 226

M94, 171, 178, 185, 255
M95, 170, 179, 186, 243, 244
M96, 170, 179, 186, 243, 244
M97, 175, 179, 191, 247
M103, 371
M104, 170, 179, 192, 271
M105, 170, 179, 187, 243
M110, 359
Makropoulos, Konstantinos, 74
Markarian's Chain, 171, 180, 192, 267, 269
Matsumoto, Tatsuro, 123
Méchain, Pierre, 150
Melotte15, 173, 272
Melotte20, 173, 176, 188, 385
Melotte25, 173, 176, 190, 197
Melotte111, 174, 177, 184, 259
Melotte186, 174, 177, 187, 299
Melotte, Philbert Jaques, 150
Messier, Charles, 150, 200, 213, 248, 314
Messier marathon, 135, 381
Mira, 175, 176, 184, 380
Mirror mounts, 110–111, 117, 131
Miyauchi, 12, 131
Mizon, Bob, xi
Monopod, 4, 47, 48, 103–104, 106, 117
Moore, Patrick Caldwell-, 150
Mu Cephei, 346
Multiple stars, 172–173, 211, 280
Muscleman Cluster, The, 173, 177, 183, 382, 383

N
Nason, Gordon, 60
Neckpod, 104–106
NGC 55, 170, 178, 190, 353
NGC 104, 171, 178, 191, 357
NGC 205, 359
NGC 221, 359
NGC 224, 359
NGC 225, 173, 177, 183, 368
NGC 250, 170, 178, 184, 361
NGC 256, 170, 178, 190, 353, 354, 361
NGC 293, 171, 178, 190, 354
NGC 297, 170, 176, 191, 358
NGC 305, 170, 179, 190, 355
NGC 436, 173, 179, 183, 369
NGC 457, 173, 179, 183, 369, 370
NGC 598, 3, 170, 176, 191, 377
NGC 654, 173, 177, 183, 372
NGC 1027, 173, 177, 183, 374
NGC 1039, 173, 176, 188, 384
NGC 1232, 170, 179, 186, 379
NGC 1342, 173, 177, 188, 386

Index 433

NGC 1499, 170, 177, 188, 194
NGC 1502, 389
NGC 1528, 173, 177, 188, 195
NGC 1535, 175, 179, 186, 196
NGC 1545, 173, 179, 188, 195
NGC 1647, 173, 177, 191, 197–199
NGC 1904, 202
NGC 1912, 173, 178, 182, 204
NGC 1952, 175, 179, 191, 200
NGC 1960, 173, 178, 182, 204
NGC 1973, 175, 179, 188, 208
NGC 1975, 175, 179, 188, 208
NGC 1976, 170, 176, 188, 208
NGC 1977, 175, 179, 188, 208
NGC 1980, 173, 176, 188, 208
NGC 1982, 170, 176, 188, 208
NGC 2024, 170, 178, 188, 212
NGC 2070, 170, 179, 185, 205
NGC 2099, 173, 178, 182, 204
NGC 2158, 214
NGC 2168, 173, 176, 186, 214
NGC 2239, 215
NGC 2244, 173, 178, 187, 215
NGC 2264, 173, 178, 187, 216
NGC 2287, 173, 176, 183, 219
NGC 2323, 173, 176, 187, 217
NGC 2353, 173, 179, 187, 218
NGC 2362, 174, 179, 183, 220
NGC 2403, 170, 179, 182, 222
NGC 2422, 174, 176, 189, 221
NGC 2437, 174, 176, 189, 221
NGC 2451, 174, 176, 189, 227, 228
NGC 2477, 174, 179, 189, 228
NGC 2516, 174, 179, 183, 223
NGC 2539, 174, 179, 189, 224
NGC 2546, 174, 179, 189, 229
NGC 2547, 174, 179, 192, 224
NGC 2632, 174, 177, 182, 232
NGC 2682, 174, 178, 182, 233
NGC 3031, 170, 179, 191, 235
NGC 3034, 170, 179, 191, 235
NGC 3114, 174, 177, 183, 238
NGC 3115, 170, 179, 190, 239
NGC 3228, 173, 178, 192, 356
NGC 3242, 175, 179, 186, 240
NGC 3351, 170, 179, 186, 243
NGC 3368, 170, 179, 186, 243
NGC 3371, 243
NGC 3372, 170, 177, 183, 242
NGC 3373, 243
NGC 3379, 170, 179, 187, 243
NGC 3521, 170, 179, 187, 244
NGC 3607, 171, 179, 187, 245
NGC 3623, 171, 179, 187, 246

NGC 3627, 171, 179, 187, 246
NGC 3628, 171, 179, 187, 246
NGC 3766, 174, 179, 183, 250
NGC 4361, 175, 180, 184, 249
NGC 4372, 171, 180, 187, 264, 265
NGC 4374, 171, 180, 192, 267
NGC 4406, 171, 180, 192
NGC 4438, 171, 180, 192
NGC 4459, 171, 180, 192
NGC 4473, 171, 180, 192
NGC 4477, 171, 180, 192
NGC 4559, 171, 180, 184, 260, 261
NGC 4565, 171, 180, 184, 261
NGC 4631, 171, 180, 185, 253
NGC 4656, 171, 180, 185, 253
NGC 4755, 174, 177, 185, 266
NGC 4833, 171, 180, 187, 265
NGC 5128, 171, 180, 183, 274
NGC 5139, 172, 177, 183, 275
NGC 5194, 171, 180, 185, 257
NGC 5195, 257
NGC 5272, 172, 178, 185, 258
NGC 5904, 172, 180, 190, 281
NGC 5907, 171, 180, 186, 277, 278
NGC 6025, 174, 180, 191
NGC 6067, 174, 180, 187, 285
NGC 6121, 172, 180, 189, 286
NGC 6205, 172, 177, 186, 302
NGC 6231, 174, 177, 189, 287
NGC 6322, 174, 180, 190, 288
NGC 6341, 172, 180, 186, 303
NGC 6397, 172, 180, 182, 304
NGC 6405, 174, 177, 190, 289
NGC 6475, 174, 177, 190, 290
NGC 6494, 174, 178, 189, 306
NGC 6496, 172, 180, 184, 305
NGC 6514, 170, 180, 189, 307, 308
NGC 6523, 170, 177, 189, 307, 308
NGC 6530, 174, 177, 189, 308
NGC 6541, 172, 180, 184, 305
NGC 6572, 175, 180, 187, 300
NGC 6584, 172, 180, 191, 318
NGC 6603, 174, 177, 189, 309
NGC 6605, 310, 311
NGC 6611, 174, 180, 190, 319
NGC 6618, 170, 180, 189, 311
NGC 6633, 174, 180, 187, 301, 320
NGC 6656, 172, 178, 189, 314
NGC 6705, 174, 177, 190, 323
NGC 6709, 174, 181, 326
NGC 6712, 172, 181, 190, 324
NGC 6723, 172, 181, 189, 316
NGC 6738, 174, 181, 327
NGC 6752, 172, 181, 188, 325

NGC 6781, 175, 181, 328
NGC 6838, 172, 189, 333
NGC 6853, 175, 177, 192, 331
NGC 6934, 172, 185, 341
NGC 6960, 175, 185, 337
NGC 6992, 175, 185, 337
NGC 6995, 185
NGC 7000, 6, 170, 177, 185, 337
NGC 7078, 172, 177, 188, 342
NGC 7209, 174, 178, 186, 347
NGC 7235, 174, 178, 184, 350
NGC 7243, 175, 178, 186, 349
NGC 7293, 175, 181, 352
NGC 7510, 175, 178, 184, 351
NGC 7686, 175, 178, 181, 364
NGC 7789, 174, 178, 183, 367
Nitrogen filling, 21, 54, 58
North American Nebula, 6, 170, 177, 185, 338
Nutrition, 141, 143

O

Objective lens, 9, 11, 15, 16, 22, 27, 29, 31, 36, 40, 44, 50, 68, 69, 71, 79, 80, 83, 84, 91, 93, 132
Observing chairs, 47, 103, 109, 113–116, 141
O-III filter, 132, 240, 247, 249, 251, 328, 337, 352
Omega Centauri, 275
Omega Nebula, The, 170, 180, 189, 311
Omicron Ceti, 175, 176, 184, 380
Omicron Velorum Cluster, 174, 177, 192, 231
Open clusters, 6, 48, 49, 144, 149, 150, 173–175, 195, 197–199, 204, 207, 210, 214–221, 223–233, 238, 241, 250, 251, 259, 266, 284–290, 297, 299, 301, 306, 308–311, 313, 320, 322, 323, 326, 327, 333, 335, 340, 345, 348–351, 356, 360, 364–376, 382–386, 388, 389
OptiClean, 81
Owl Cluster, The, 173, 179, 183, 369, 370

P

Parallelogram mounts, 4, 103, 111–113, 117, 118
Parsonstown Leviathan, The, 257
Patriarca, Larry, xi
PDA. *See* Personal digital assistant (PDA)
Pechan roof prism, 19, 21
Personal digital assistant (PDA), 136, 138
Petzval, 11
Phase coating, 20
Pinwheel Galaxy, The, 3, 6, 170, 176, 191, 377

Planetarium, 81, 125, 136–139
Planetary nebulae, 132, 149, 175
PleiadAtlas, 136, 139
Pleiades, The, 173, 176, 190, 198, 388
Pocket stars, 138
Polaris, 235, 373, 387
Porro prism, 9, 12–15, 20, 21, 36–38, 53, 54, 65, 83, 96, 100–102
Porro type-2 prism, 14, 15
Praesepe, 174, 177, 182, 232
Prism adjustment, 92–93
Prisms, 9, 12–27, 35, 40, 43, 44, 57, 58, 65, 67, 77, 83, 87, 88, 91–93, 96, 100
Psi-1 Piscium, 172, 179, 188, 362
Ptolemy's Cluster, 174, 177, 190, 290
Push to, 125, 137

Q

Quality control, 27, 43, 44, 47, 57, 58, 64

R

Rain guards, 78–80
Rank, David, 11
Red Giant, 149, 346, 380
Reflection nebulae, 149–150, 175
Reflex finders, 130–131
Relative brightness, 44–45, 53
Rhomboid prism, 22
Rho Ophiuchi, 172, 180, 187, 291
Rifle Sling hold, 98
RKE, 11
R Leporis, 175, 177, 187, 201
Roof prism, 9, 12, 17, 19–21, 36–38, 53, 55, 67, 97, 100, 102
Rosette Nebula, The, 215
Rosse, Lord, 257
Running Chicken, The, 174, 179, 183, 251
RV Boötis, 176, 180, 182, 279
RW Boötis, 279

S

Sayre, Bruce, xi, 124
Schmidt roof prism, 17, 20, 21
Schott, A.G., 17, 44
Scutum Star Cloud, 323
Sechii, Angelo, 254
Semi penta prism, 17, 19, 21
9 Sextantis, 172, 179, 190, 234
Seyfried, J.W., 61, 94
Shoes, 143
Sighting tubes, 129, 130

Sigma Orionis, 172, 176, 188, 211
Simmons, Craig, 115
Sirius, 219, 221, 242
2Sky, 136
Sky Vector, 125
Small Magellanic Cloud, 170, 176, 191, 357, 358
Smartphone, 136
Software, 135–137, 139
Solar filter, 132, 139
Sombrero Galaxy, The, 170, 179, 192, 271
Spherical aberration, 11, 29–31, 43, 49, 69, 70
Spindle Galaxy, The, 170, 179, 190, 239
Star chair, 115–117
Step, 40, 61, 72, 90
Stereopsis, 5, 58, 209
Stock 2, 173, 177, 183, 374, 382, 383
Stock 12, 175, 365
Stock 23, 173, 375
Storage, 78–79, 82, 113, 117, 138–139
Strange, David, xi
Struve 2809, 343
Struve, Friedrich Georg Wilhelm von, 150
Summation, 5, 58
Supernova remnants, 175, 200, 251, 337
Swan Nebula, The, 170, 180, 189, 311

T
Takahashi, 8, 35, 47, 121
Tarantula Nebula, The, 170, 179, 185, 205
Teeter, Rob, xi
Theta Orionis, 208
Theta Pictoris, 172, 179, 188, 206
Theta Serpentis, 172, 181, 190, 320, 321
Tonkin, S., 63
Torches, 113, 137–139
Trapezium, The, 208, 209
Triangular Arm Brace, 96–98
Trifid Nebula, The, 170, 180, 189

Trigger-grip ball-head, 4, 47, 103, 105, 117
Tripod bush, 66, 84, 100, 101
Tripod heads, 55, 100, 102, 105, 107, 108
Tripods, 4, 47, 55, 100, 103, 107–113, 116, 117, 139
True Field of View, 49, 74
47 Tucanae, 171, 178, 191, 357, 358
Twilight Index, 45
Twilight Performance Factor, 45

U
UHC filter, 132, 216, 240, 247, 249, 251, 311, 319, 328, 337, 345, 352, 374

V
Variable stars, viii, 3, 175–176, 201, 279, 380, 384
Vari-angle prism, 22, 23
Vee and blade sight, 129, 130
Veil Nebula, The, 175, 185, 337
Vertical misalignment, 40
Vignetting, 16, 35, 43, 49, 67, 71
Visibility factor, 45
Vixen, 121

W
Whirlpool Galaxy, The, 171, 180, 185, 257

Y
Y Cvn, 175, 177, 185, 254

Z
Zarenski, E.D., xi, 46
Zeiss body, 84
Zeta Piscium, 172, 179, 188, 363
Zoom binoculars, 60

Made in the USA
Monee, IL
18 December 2019